AF869586

WITHDRAWN
FROM
WADHAM COLLEGE
LIBRARY
DATE
LIBRARIAN

Studies in Modern Chemistry

Advanced courses in chemistry are changing rapidly in both structure and content. The changes have led to a demand for up-to-date books that present recent developments clearly and concisely. This series is meant to provide advanced students with books that will bridge the gap between the standard textbook and the research paper. The books should also be useful to a chemist who requires a survey of current work outside his own field of research. Mathematical treatment has been kept as simple as is consistent with clear understanding of the subject.
Careful selection of authors actively engaged in research in each field, together with the guidance of four experienced editors, has ensured that each book ideally suits the needs of persons seeking a comprehensible and modern treatment of rapidly developing areas of chemistry.

William C. Agosta, The Rockefeller University
R. S. Nyholm, FRS, University College London

Consulting Editors

Academic editor for this volume

Alan Carrington, University of Southampton

Studies in Modern Chemistry

R. L. M. Allen
Colour Chemistry

A. Gilbert and R. B. Cundall
Photochemistry

T. L. Gilchrist and C. W. Rees
Carbenes, Nitrenes, and Arynes

S. F. A. Kettle
Coordination Compounds

Ruth M. Lynden-Bell and Robin K. Harris
Nuclear Magnetic Resonance Spectroscopy

A. G. Maddock
Lanthanides and Actinides

E. S. Swinbourne
Analysis of Kinetic Data

T. C. Waddington
Non-aqueous Solvents

K. Wade
Electron Deficient Compounds

Nuclear magnetic resonance spectroscopy

Ruth M. Lynden-Bell
Lecturer in the Department of Chemistry
University of Sussex
Robin K. Harris
Lecturer in the School of Chemical Sciences
University of East Anglia

WADHAM COLLEGE
LIBRARY.

Nelson

Thomas Nelson and Sons Ltd
36 Park Street London W1
P.O. Box 336 Apapa Lagos
P.O. Box 25012 Nairobi
P.O. Box 21149 Dar es Salaam
P.O. Box 2187 Accra
77 Coffee Street San Fernando Trinidad
Thomas Nelson (Australia) Ltd
597 Little Collins Street Melbourne 3000
Thomas Nelson and Sons (South Africa) (Proprietary) Ltd
51 Commissioner Street Johannesburg
Thomas Nelson and Sons (Canada) Ltd
81 Curlew Drive Don Mills Ontario

First published in Great Britain 1969
Copyright © Ruth M. Lynden-Bell and Robin K. Harris 1969

Illustrations by Colin Rattray

17 771710 6 (Paper)
17 761711 X (Boards)

Printed in Northern Ireland at The Universities Press, Belfast

Contents

Preface

Since the first nuclear magnetic resonance experiments were carried out shortly after the Second World War, this branch of spectroscopy has developed at a rapid and accelerating pace. Most large chemical laboratories, both industrial and academic, possess at least one high-resolution NMR spectrometer, and the use of the technique is commonplace. No university undergraduate chemistry course is complete without a discussion of the basic principles, and most courses include practical experience in NMR. There are several books or reviews on the subject, but unfortunately these are written at the research level, or else deal mostly with applications. General textbooks on spectroscopy usually include a chapter on the topic but, we feel, do not give adequate treatment. There seemed to be a need for a short monograph dealing with the physico-chemical basis of the subject. The present text was produced to fill that need. It is intended to cover the topic adequately for students taking Bachelor's and Master's degree courses in chemistry. It is, perhaps, particularly suitable for those whose main interest lies in the fields of physical chemistry, chemical physics, or theoretical chemistry. The material presented covers the courses given currently at the Universities of East Anglia and Sussex. Clearly some aspects have had to be omitted or treated very briefly, but we have aimed to deal with all the fundamentals of high-resolution NMR.

The book is arranged so that students at different levels can benefit from studying it. The first chapter contains basic material which all chemistry undergraduates should know. The following chapters can be read independently of each other. Chapter 2 develops the quantum-mechanical approach to spectral analysis which is necessary for all but the simplest spectra and which incidentally illustrates many of the principles of quantum mechanics. Chapter 3 is concerned with the origin of chemical shifts and coupling constants on the molecular level and their relation to molecular structure. The last chapter deals with some further spectroscopic phenomena, the first part with the dependence of line shapes on relaxation and rate processes and the second part with double resonance experiments and their uses.

We have provided problems at the end of each chapter to illustrate the text and in some cases to introduce further ideas. At the end of the book is a list of suggestions for further reading on many of the topics mentioned or discussed, but we consider it inappropriate to provide detailed references to the research literature in a book of this kind.

Magnetic resonance is an appropriate area for the early introduction of SI or MKSA units, which are now internationally recommended for adoption in all scientific work, as the subject develops logically from physics courses on electricity and magnetism, where MKSA units are normally used. Students are at present often taught inconsistently in related physics and chemistry courses. In this book we have used SI (MKSA) units for numerical quantities and written electric and magnetic equations in a rationalized form. A brief section on units follows this preface.

We are grateful to those of our colleagues who read part or all of the manuscript and who provided spectra.

Ruth M. Lynden-Bell
Robin K. Harris

Units

At the outset it was decided to use the Système International (SI)—often known in the U.S.A. as the MKSA system—for the units of physico-chemical quantities, since this has been accepted by the international scientific authorities as the convention for all future work. However, this system is unfamiliar to many chemists who are used to the old centimetre-gramme-second (CGS) units; to such people we recommend the booklet by M. L. McGlashan entitled 'Physico-Chemical Quantities and Units' (*Royal Institute of Chemistry Monographs for Teachers*, Number 15, 1968). This should also be read by those wishing to study NMR in greater depth since most books and research papers are based on CGS units. A summary of SI units and notation is given here following this discussion. One particular point needs to be made at this stage. Magnetic phenomena may be discussed in terms of two field vectors, known as the magnetic field intensity, **H**, and the magnetic induction field (or magnetic flux density), **B**. Observable physical properties depend directly on **B** rather than **H** (which is a convenient auxiliary vector for calculations using Ampère's circuital theorem). However, the chemist (and particularly the NMR spectroscopist) has been accustomed, incorrectly, to write **H** instead of **B**. This is tolerable when using electromagnetic units (e.m.u.) since **H** and **B** then have the same magnitude; it is not permissible in SI since **H** and **B** differ by $4\pi \times 10^{-7}$. Consequently, we consistently use **B**, which for simplicity we refer to as the magnetic field.

The SI units really incorporate four modifications of the CGS system:

(1) The basic quantities of length, mass, and time are the metre, kilogramme, and second. In addition the ampere is chosen as the basic unit of electrical current. All other quantities are expressed in units coherent with these four (plus a handful of additional basic units which will not concern us here). For instance, energy is expressed in joules and power in watts. On this scheme most electrical and magnetic quantities are in familiar units (coulombs, ohms, volts, etc.). The only striking exception is the unit of magnetic induction field, which in SI is the tesla (10^4 times the more familiar electromagnetic unit, the gauss).

(2) The SI treats electrical and magnetic units coherently, whereas former practice has been to use separate electrostatic units (e.s.u.)

WADHAM COLLEGE
LIBRARY.

and electromagnetic units (e.m.u.) in different problems. Whenever e.s.u. and e.m.u. were used in the *same* problem the velocity of light, c, appeared in equations. This was frequently inconvenient to remember. The quantity c does *not* appear in the analogous SI equations. Thus the magnetic moment, μ, due to a current i in a loop of area A may be written $\mu(\text{e.m.u.}) = i(\text{e.s.u.})A/c$ or $\mu(\text{e.m.u.}) = i(\text{e.m.u.})A$, whereas in SI it is invariably $\mu = iA$.

(3) The e.s.u.–e.m.u. system does not use a separate 'dimension' for electrical and magnetic properties whereas the SI system does. The permittivity and permeability of free space (ε_0 and Ξ_0 respectively) have dimensions and often appear explicitly in SI equations where they are sometimes omitted (because they have unit magnitude) in e.s.u. and e.m.u. equations. Another result is that the units of certain quantities are not always related in the simple manner one would suppose from (1) above. Thus, reduced coupling constants (see Section 3-16) are in cm^{-3} in the CGS system, but contain magnetic quantities implicitly. In SI the magnetic quantities are explicit and the units are newton ampere^2 metre^{-3}, not simply metre^{-3}.

(4) SI equations are written 'rationalized' whereas CGS equations are usually 'unrationalized.' Rationalized equations have factors of 2π and 4π where expected on geometrical grounds, but not otherwise, whereas the reverse is true for unrationalized equations. One result from this is that magnetic susceptibilities in SI are 4π times those in e.m.u.

The use of the SI system has two consequences in this book. Firstly, some equations differ from those printed elsewhere by factors involving (i) c, (ii) Ξ_0, and/or (iii) 4π. Secondly, where numerical values are given these may be of unfamiliar magnitudes and in unfamiliar units. Fortunately the quantities most often used by nuclear resonance spectroscopists (frequencies, coupling constants, and chemical shifts) are the same in CGS and SI units, and other quantities are often quoted for comparative purposes only. These differences may seem somewhat daunting at first sight (though many electricity and magnetism textbooks already use MKSA units), but the advantages of using a completely coherent set of units will soon be felt, and we believe that the task of teaching and learning scientific subjects will be made easier. Moreover, all completely algebraic equations are unitless because the symbols for quantities include the units. It is, of course, necessary when substituting numerical values in the equations to ensure that choice of units is *consistent*.

List of SI units and related quantities

Property	*Symbol*	*SI unit*	*Symbol of unit*	*Equivalent SI units*	*Equivalents in other systems*
Length	ℓ, r	metre	m	–	10^2 centimetres
Mass	m	kilogramme	kg	–	10^3 grammes
Time	t	second	s	–	–
Force	F	newton	N	kg m s^{-2}	10^7 dynes
Energy	U	joule	J	kg m^2 s^{-2}	10^7 ergs
Angle	θ	radian	rad	$1/2\pi$ cycles	–
Frequency	ν	hertz	Hz	–	1 cycle per second (c/s or cps)
Electric charge	q	coulomb	C	A s	$2{\cdot}9979 \times 10^9$ e.s.u.
Electric current	i	ampere	A	–	(1/10) e.m.u.
Magnetic induction field	B	tesla	T	kg s^{-2} A^{-1}	10^4 gauss (e.m.u.)
Magnetic field intensity	H	ampere metre^{-1}	–	m^{-1} A	$4\pi \times 10^{-3}$ oersted
Permeability	Ξ	newton ampere^{-2}	–	kg m s^{-2} A^{-2}	–
Magnetic dipole moment	μ	ampere metre^{-2}	–	m^{-2} A	10^3 e.m.u. sec^{-1}
Magnetogyric ratio	γ	radian tesla^{-1} second^{-1}	–	rad kg s^{-3} A^{-1}	$(1/10^4)$ rad gauss^{-1} sec^{-1}
Reduced coupling constant	K	newton ampere^2 metre^{-3}	–	kg m^{-2} s^{-2} A^2	10 cm^{-3}

Fundamental constants

Constant	Symbol	Value	Unit	
			SI	CGS
Elementary charge	e	1·6021	$\times 10^{-19}$ C	$\times 10^{-20}$ e.m.u.
		4·8030	–	$\times 10^{-10}$ e.s.u.
Electron rest mass	m_e	9·1091	$\times 10^{-31}$ kg	$\times 10^{-28}$ g
Proton rest mass	m_p	1·6725	$\times 10^{-27}$ kg	$\times 10^{-24}$ g
Planck constant	h	6·6256	$\times 10^{-34}$ J s	$\times 10^{-27}$ erg s
	$\hbar = h/2\pi$	1·0545	$\times 10^{-34}$ J s	$\times 10^{-27}$ erg s
Charge-to-mass ratio for electron	e/m_e	1·75880	$\times 10^{11}$ C kg^{-1}	$\times 10^{7}$ e.m.u.
Permeability constant	Ξ_o	4π	$\times 10^{-7}$ kg m s^{-2} A^{-2}	–
		1	–	$\times$ 1 e.m.u.
Free electron Landé splitting factor	g_s	2·0023	–	–
Magnetogyric ratio of proton	γ_p	2·67519	$\times 10^{8}$ rad s^{-1} T^{-1}	$\times 10^{4}$ rad s^{-1} gauss^{-1}
	$\gamma_p/2\pi$	4·25770	$\times 10^{7}$ Hz T^{-1}	$\times 10^{3}$ s^{-1} gauss^{-1}
Bohr magneton	μ_B	9·2732	$\times 10^{-24}$ J T^{-1}	$\times 10^{-21}$ erg gauss^{-1}
Nuclear magneton	μ_N	5·0505	$\times 10^{-27}$ J T^{-1}	$\times 10^{-24}$ erg gauss^{-1}
Boltzmann constant	k	1·3805	$\times 10^{-23}$ J K^{-1}	$\times 10^{-10}$ erg K^{-1}
Avogadro constant	N	6·0225	$\times 10^{23}$ mole^{-1}	$\times 10^{23}$ mole^{-1}

1 Basic principles

1-1 Introduction

Spectroscopy may be defined as the interaction between matter and electromagnetic radiation such that energy is absorbed or emitted according to the Bohr frequency condition, $\Delta E = h\nu$, where ΔE is the energy difference (normally quantized) between the initial and final states of the matter, h is Planck's constant, and ν is the frequency of the electromagnetic radiation. It proves convenient to sub-divide the subject according to the nature of the energy levels considered. This classification usually correlates with the region of the spectrum (i.e. the range of frequencies) used. Altogether, spectroscopy is the most powerful tool the chemist has for investigating molecular structure and molecular processes. This statement is true regardless of whether the chemist's interest is organic, inorganic, physical, or theoretical.

This book sets out to discuss the physico-theoretical basis of that branch of spectroscopy known as nuclear magnetic resonance (NMR). NMR involves the magnetic energy of nuclei when they are placed in a magnetic field, and the transitions occur in the radio wave region of the spectrum. The first experiments were carried out in 1945, but useful chemical applications became possible only after the discovery of the chemical shift effect in 1949. From modest beginnings the subject developed very rapidly, especially after the appearance of the first commercial NMR spectrometer in the mid-nineteen fifties. Today the subject has expanded so that it is of equal importance with the older-established vibrational (infrared) and electronic (ultraviolet) branches of spectroscopy. Indeed in many areas NMR is superior to IR and UV, but the cost of instrumentation is usually greater.

All forms of spectroscopy give spectra that may be described in terms of the frequency, intensity and shape of spectral lines or bands. These observable properties depend on molecular parameters of the system, which for NMR are found to be the shielding constants and coupling constants of the nuclei, and the lifetimes of the energy levels. It is these parameters which are of fundamental physico-chemical importance. The NMR spectroscopist will be concerned in (a) obtaining spectra, (b) evaluating the nuclear magnetic parameters from the spectra, (c) interpreting or predicting these parameters using physical or theoretical models, and (d) applying the knowledge gained to problems of chemical interest such as structure determination, evaluation of kinetic or thermodynamic data, etc. In this book we shall be concerned with all four aspects.

We shall deal almost exclusively with NMR of liquid or solution systems. Studies in the gas phase are uncommon, partly because of the rather low sensitivity of the NMR technique (vapour pressures of the order of one atmosphere would be needed) and partly because the resonance lines are usually rather broader (of the order of a few Hz) than for liquids, since collision narrowing does not always occur efficiently for gases at atmospheric pressure. NMR studies of solids do yield some useful chemical information about crystal structure, such as interproton distances, but both theory and experimental techniques differ from those used for fluids. The spectra from solids usually contain very broad bands (of the order of tens of kHz), due to the direct effects of the interaction of nuclear spin dipoles in fixed orientations, whilst those for fluids are very narrow, since molecular tumbling averages the direct effects to zero. Experimental techniques are being developed to narrow the lines for solids. Consequently, instruments with very high resolving power are needed for the study of fluids, and this branch of NMR spectroscopy is often known as high-resolution NMR, whereas work with solids is generally referred to as broad-line NMR. Studies of solids do not in general yield values for the chemical shift and coupling constant parameters with which we shall be much concerned. The term 'high-resolution' implies that it is the spectrum itself (i.e. transition frequencies and intensities, including band shapes) that is of importance. Other types of NMR experiment on fluids, such as 'spin echo' work, which is concerned with obtaining the lifetimes of energy levels, will only be mentioned briefly in passing.

This first chapter deals with the basis of the subject of NMR, explains the concepts involved, introduces the notation and instrumentation used in high-resolution work, and sets the scene for the more detailed discussion in the remaining chapters. The treatment of energy level populations, absorption line width, relaxation, and of the effects of chemical equilibria on the spectra is postponed to Chapter 4.

1-2 Quantization of angular momentum

One of the fundamental postulates of modern chemical physics is that the total angular momentum of any isolated particle cannot have any arbitrary magnitude but may only take certain discrete values. Angular momentum is said to be quantized, and its magnitude, P, can be specified in terms of a quantum number, say R, by means of Eqn. (1-1).

$$P = \hbar[R(R + 1)]^{1/2} \tag{1-1}$$

The constant $\hbar$ is $h/2\pi$, where h is known as Planck's constant; R is either integral or half-integral. Frequently, the various possible modes of motion (rotation, vibration, etc.) that contribute to the total angular momentum may be treated independently, and each assigned a separate quantum number. This quantum number specifies [by means of an equation analogous to (1-1)] the angular momentum attributed to the particular mode of motion considered. In a similar fashion the total energy of the particle may be considered as the sum of individual contributions due to the various possible modes of motion. The total wave function describing the particle may then be written as a *product* of separate wave functions for each type of motion. The quantization of angular momentum implies quantization of energy. Each energy state of the particle may be specified by the appropriate quantum numbers and has a stationary-state wave function (eigenfunction). In some instances, states with different eigenfunctions may have the same energy, and they are then said to be degenerate.

Angular momentum is actually a vector property, so that for a full description direction has to be specified as well as magnitude. In quantum-mechanical terms this is done by using another quantum number, m_R say, such that the magnitude of the component of angular momentum, P_z, along the direction considered (z), is given by Eqn. (1-2).

$$P_z = \hbar m_R \tag{1-2}$$

Quantum numbers specifying magnitude must, of course, be positive. Since the component cannot be greater than the vector itself, $|m_R|$ must be less than or equal to R, but m_R can be either positive or negative. In fact, quantum mechanics show that the allowed values of m_R are as in Eqn. (1-3); there are thus $2R + 1$ possible values of m_R.

$$m_R = R, R - 1, R - 2 \ldots - R \tag{1-3}$$

Figure 1-1 illustrates the general situation for a quantized angular momentum vector.

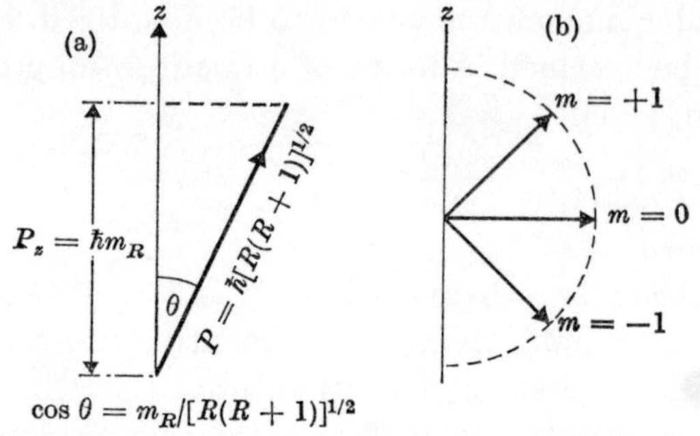

Fig. 1-1 Angular momentum vectors: (a) the general case, showing the use of quantum numbers R and m_R, (b) the possible orientations for an angular momentum vector of quantum number $R = 1$. Note that R and m_R do not specify direction in the xy plane at all.

1-3 Electron and nuclear spin

The particular type of angular momentum that this book will be dealing with is that of nuclear spin. Transitions between nuclear spin energy levels give rise to the phenomenon of NMR; the corresponding property of electron spin leads to electron spin resonance (ESR). The concept of electron spin is probably familiar to chemistry students at an early stage, since the structure of the periodic table can be explained on the basis of four electronic quantum numbers, n, ℓ, m_ℓ and m_s, the last-mentioned being the directional spin quantum number. The postulate that electrons possessed angular momentum which could be attributed to a spinning motion was first made in 1925, in order to explain both the fine structure of bands observed in atomic spectra, and the effect of a magnetic field on such spectra (the anomalous Zeeman effect). Electron spin also appears as a natural result from relativistic wave mechanics. The spin quantum number of a single electron is $s = \frac{1}{2}$ and the associated angular momentum is $\hbar\sqrt{3}/2$ [see Eqn. (1-1)]. The possible values of the directional quantum number, m_s, are therefore $+\frac{1}{2}$ and $-\frac{1}{2}$. This is reflected in one of the ways of expressing Pauli's exclusion principle, namely that only two electrons ($m_s = \pm\frac{1}{2}$) may occupy an electronic orbital specified by the quantum numbers, n, ℓ, and m_ℓ.

The existence of nuclear spin was first suggested by Pauli in 1924, a year before the idea of electron spin was mooted, in order to explain further splittings in atomic spectra (hyperfine structure). These splittings demonstrate that nuclei of different elements (and different isotopes of the same element) differ in spin angular momentum. This fact may be traced to details of the internal construction of nuclei. The proton has a spin quantum number, I, of $\frac{1}{2}$, as does the neutron. For nuclei other than ^1H the spin angular momenta of the individual nucleons couple (together with their orbital-type angular momenta) to give the observed total. This is characterized by a quantum number, I, which may be integral or half-integral or zero. Excited states of the nucleus with different values of I exist; these are important for the Mössbauer effect, but the energy differences between the possible states is so huge that only the ground state need be considered here. Certain systematic features may be noted as follows:

(a) Nuclei with an odd mass number have half-integral spin.

(b) Nuclei with an even mass number and an even charge number have zero spin.

(c) Nuclei with an even mass number but an odd charge number have integral spin.

Table 1-1 lists values of the nuclear spin quantum number for some of the more common species. Since NMR depends on the existence of nuclear spin, nuclei with $I = 0$ do not show any direct effects and for most purposes may be ignored. It will be noted from Table 1-1 that such nuclei include ^{12}C and ^{16}O. Since these isotopes account for almost the whole of naturally occurring carbon and oxygen nuclei, the NMR spectra of many quite complex organic chemicals are remarkably simple and easy to interpret, being attributable to hydrogen nuclei (protons) alone. It is this fact that makes NMR in many cases a more convenient tool for structure determination than infrared or ultraviolet spectroscopy.

The direction of a spin angular momentum vector is conventionally defined such that the spinning body, viewed along the vector, twists in the same sense as a right-handed screw. The same convention is applied to the vector $\boldsymbol{\omega}$ describing the angular velocity of the rotation. The magnitude of the spin angular momentum is given by the principles of Section 1-2 as

$$P = \hbar[I(I+1)]^{1/2} \tag{1-4}$$

Nuclei with spin quantum number greater than $\frac{1}{2}$ also possess electric quadrupole moments. It will be shown in Chapter 4 that this results in lifetimes for their nuclear magnetic states which are normally much shorter than those for spin $\frac{1}{2}$ nuclei. The NMR absorption lines of $I > \frac{1}{2}$ nuclei are therefore broad (by virtue of the uncertainty principle) and not very readily studied. For many purposes such nuclei may be regarded as effectively non-magnetic. For most of this book we shall be considering only spin $\frac{1}{2}$ nuclei, though some of the equations and the majority of the concepts apply equally to spin $> \frac{1}{2}$ nuclei. Some specific properties of such nuclei will be considered in Chapter 4.

1-4 Nuclear magnetic moments

Any motion of a charged body has an associated magnetic field; on a macroscopic scale an electrical current, which is due to motion of electrons along a conductor, produces a magnetic field. Current travelling in a loop has an associated magnetic dipole moment. This phenomenon also occurs on an atomic scale, for whenever electrons or nuclei possess angular momentum there is a magnetic moment. Since angular momentum is quantized on this scale, so are magnetic moments. The magnitude of such moments is most easily discussed from the classical standpoint by considering orbital electronic motion. Suppose an electron is travelling in an orbital at an angular velocity ω. Such motion is equivalent to an electrical current in the opposite direction of magnitude $i = e\omega/2\pi$, where e is the magnitude of the charge on the electron. The orbital angular momentum, denoted P, is $m_e r^2 \omega$ where m_e is the mass of the electron and r its distance from the nucleus. Thus, the current is

$$i = e\omega/2\pi = eP/2\pi m_e r^2 \tag{1-5}$$

The magnetic moment μ generated by such motion is given in electromagnetic theory by $\mu = Ai$, where A is the area marked out by the orbital. If it is supposed, for simplicity, that the electron moves in a circular orbit, $A = \pi r^2$, and substitution in Eqn. (1-5) gives

$$\boldsymbol{\mu} = -(e/2m_e)\mathbf{P} \tag{1-6}$$

This equation illustrates the fact that μ and P are anti-parallel vectors for electrons. As has been stated above, on the atomic scale angular momentum is quantized in units of $\hbar$. It follows that electron magnetic moments are quantized in units of $(eh/4\pi m_e)$. This unit, which will be written μ_B, is an important quantity, known as the Bohr magneton. It has a value of $9{\cdot}2732 \times 10^{-24}$ JT^{-1} $\equiv 9{\cdot}2732 \times 10^{-21}$ erg gauss^{-1}. When spin motion is involved, this simple approach does not reproduce quantitatively the quantum-mechanical expression relating magnetic moment and angular momentum. It is convenient to introduce a factor g (known as the Landé splitting factor, or simply the g-factor) to take account of this fact. Thus, we write

$$\boldsymbol{\mu} = -g(e/2m_e)\mathbf{P} = -g\mu_B\mathbf{P}/\hbar \tag{1-7}$$

It is important to note that the observed values of the g-factor in various situations can be reproduced by a full quantum-mechanical treatment. Equation (1-7) can also be used for coupled spin–orbital motion if appropriate values of g are used.

Analogous equations may be used for the spin motion of nuclei. The constant of proportionality relating μ and P depends on the mass of the particle and on its charge. The appropriate equation for nuclei is therefore Eqn. (1-8)

$$\hat{\mu} \propto (Ze/2m_N)\hat{P} \tag{1-8}$$

where m_N is the mass of the nucleus and Z its charge number. For nuclei it proves to be convenient to use a unit of magnetic moment based on the application of Eqn. (1-8) to the proton. This unit is known as the nuclear magneton, $\mu_N = eh/4\pi m_p$, where m_p is the mass of the proton. The magnitude of μ_N is $5{\cdot}0505 \times 10^{-27}\ \mathrm{JT}^{-1} \equiv 5{\cdot}0505 \times 10^{-24}$ erg gauss^{-1}. For other nuclei the charge number and the difference in mass from that of the proton are both included in the nuclear g-factor, which also allows for quantum-mechanical effects. Thus

$$\boldsymbol{\mu} = g_N \mu_N \mathbf{P}/\hbar \tag{1-9}$$

where g_N is the *nuclear* g-factor. It should be noted that unlike Eqn. (1-7), which includes a negative sign explicitly (so that g is positive for electrons), Eqn. (1-9) allows the nuclear g-factor to be either positive or negative. It might be thought that, since all nuclei are positively charged, all nuclear g-factors must be positive. However, this is not so, and the reason for this may be traced to the structure of the nucleus. Negative nuclear g-factors imply that the spin magnetic moment is anti-parallel to the angular momentum (as for the electron); positive values indicate that $\boldsymbol{\mu}$ and $\mathbf{P}$ are parallel.*

* For protons $g = 2{\cdot}79245$ For neutrons $g = -1{\cdot}9135$ [See internal nucleon structure]. There is no suitable theory for g values!

Usually, nuclear magnetic moments are described in terms of magnetogyric ratios, γ, rather than nuclear g-factors. The value of γ is defined as the ratio of the magnetic moment to the angular momentum (taking explicit account of the sign of the relationship).

$$\boldsymbol{\mu} = \gamma \mathbf{P} \tag{1-10}$$

From Eqns. (1-9) and (1-10) it can be seen that

$$\gamma\hbar = g_N \mu_N \tag{1-11}$$

Selected values of nuclear magnetic dipole moments and magnetogyric ratios are given in Table 1-1. Use of Eqn. (1-4) with Eqns. (1-9) and (1-10) gives

$$|\mu| = g_N \mu_N [I(I+1)]^{1/2} = \gamma\hbar[I(I+1)]^{1/2} \tag{1-12}$$

Nuclear g-factors are small numbers (as are electron g-factors) since the effects of mass and charge number are partially compensatory.

Table 1-1 Spin properties for a selection of nuclei

Isotope	I	*Natural abundance* (%)	$\mu/\mu_N^{(a)}$ (In units of Bohr nuc. Magnetons.)	$\gamma^{(b)}$ (10^7 rad $T^{-1}s^{-1}$)	$Q^{(c)}$ (10^{-28} m^2)
electron	$\frac{1}{2}$	–	$-3{\cdot}184 \times 10^3$	$-1{\cdot}761 \times 10^4$	–
neutron	$\frac{1}{2}$	–	−3·313	−18·325	–
^{1}H	$\frac{1}{2}$	99·98	4·837	26·7519	–
^{2}H	1	0·016	1·212	4·106	0·0028
^{10}B	3	18·83	2·078	2·875	0·111
^{11}B	$\frac{3}{2}$	81·17	3·470	8·583	0·0355
^{12}C	0	98·89	–	–	–
^{13}C	$\frac{1}{2}$	1·11	1·216	6·726	–
^{14}N	1	99·64	0·571	1·933	0·02
^{15}N	$\frac{1}{2}$	0·36	−0·490	−2·711	–
^{16}O	0	99·76	–	–	–
^{17}O	$\frac{5}{2}$	0·037	−2·239	−3·627	−0·004
^{19}F	$\frac{1}{2}$	100	4·550	25·167	–
^{27}Al	$\frac{5}{2}$	100	4·304	6·971	0·149
^{28}Si	0	92·28	–	–	–
^{29}Si	$\frac{1}{2}$	4·67	−0·961	−5·314	–
^{30}Si	0	3·05	–	–	–
^{31}P	$\frac{1}{2}$	100	1·958	10·829	–
^{32}S	0	95·06	–	–	–
^{33}S	$\frac{3}{2}$	0·74	0·830	2·052	−0·064
^{34}S	0	4·18	–	–	–
^{35}Cl	$\frac{3}{2}$	75·4	1·060	2·621	−0·0797
^{37}Cl	$\frac{3}{2}$	24·6	0·882	2·182	−0·0621
^{116}Sn	0	14·28	–	–	–
^{117}Sn	$\frac{1}{2}$	7·67	−1·723	−9·530	–
^{118}Sn	0	23·84	–	–	–
^{119}Sn	$\frac{1}{2}$	8·68	−1·803	−9·971	–
^{120}Sn	0	32·75	–	–	–
^{198}Hg	0	10·02	–	–	–
^{199}Hg	$\frac{1}{2}$	16·92	0·865	4·783	–
^{200}Hg	0	23·10	–	–	–
^{201}Hg	$\frac{3}{2}$	13·22	−0·784	−1·938	0·5
^{202}Hg	0	29·72	–	–	–

(a) Magnetic dipole moment in nuclear magnetons (the magnitude listed is $g_N[I(I+1)]^{1/2}$).
(b) Magnetogyric ratio; in e.m.u. the numbers given are in multiples of 10^3 rad gauss^{-1} s^{-1}.
(c) The electric quadrupole moment (see Section 4-7) is eQ, where e is the charge on the proton.

Nuclear magnetic moments are, however, only ca. 10^{-3} times as big as electronic magnetic moments due to the effect of the mass to charge ratio—the value of μ_N/μ_B depends on m_e/m_p.

1-5 Nuclei in a magnetic field

Felix Block 1946-49.

In the absence of a magnetic field the energy of an isolated nucleus is independent of the quantum number m_I. This is equivalent to stating that the energy of a magnet is independent of its orientation when no magnetic field is present. However, when a magnetic field of strength B is applied in a direction defined as z, a magnetic moment $\boldsymbol{\mu}$ would have, in classical terms, an energy U given by

$$U = -\boldsymbol{\mu} \cdot \mathbf{B} = -\mu_z B \tag{1-13}$$

where μ_z is the component of $\boldsymbol{\mu}$ in the direction z. This component is, in the case of nuclear spin, defined by m_I and [using an equation

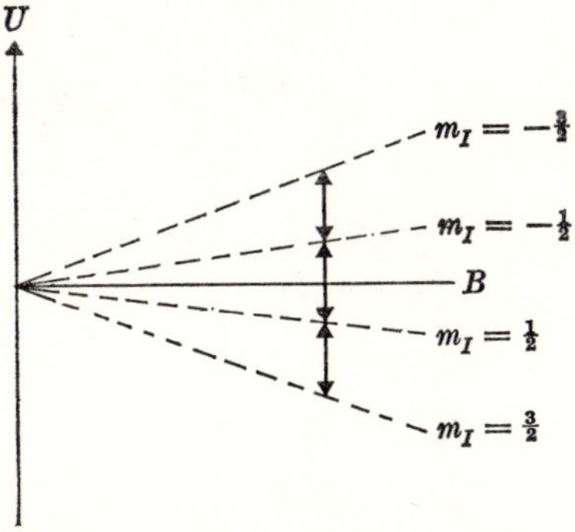

Fig. 1-2 The nuclear spin energy for a single nucleus with $I = \frac{3}{2}$ (such as ^{11}B), plotted as a function of the magnetic field B. The transitions obeying the selection rule $\Delta m_I = \pm 1$ for a particular value of B are indicated; these three transitions are degenerate for an isolated spin.

analogous to Eqn. (1-12)] given by

$$\mu_z = g_N \mu_N m_I = \gamma \hbar m_I \tag{1-14}$$

Thus Eqn. (1-13) becomes

$$U = -\gamma \hbar m_I B \tag{1-15}$$

In this situation there are $(2I + 1)$ non-degenerate energy levels (see Fig. 1-2) corresponding to the $2I + 1$ values of m_I, each separated by $|\gamma \hbar B|$. It is clear that in principle transitions may be induced between these levels by the use of appropriate electromagnetic radiation. Such transitions will be governed by the Bohr frequency condition, whereby energy is transferred when ΔU, the gap between energy levels considered, is equal to $h\nu$, where ν is the frequency of the radiation.

Thus for NMR transitions of a single nucleus

$$h\nu = |\gamma \hbar B \, \Delta m_I| \tag{1-16}$$

$$\nu = |(\gamma/2\pi) B \, \Delta m_I| \tag{1-17}$$

The selection rule governing such transitions is $\Delta m_I = \pm 1$ (see Chapter 2) so Eqn. (1-17) reduces to

$$\nu = |\gamma/2\pi|\, B \tag{1-18}$$

It may be seen that for $I > \frac{1}{2}$ the $2I$ possible transitions are all degenerate in this simple case. Normally, magnetic fields in the range 1–5 T (10–50 kgauss) are used, and from Table 1-1 it can be seen that frequencies of the order of tens of MHz are necessary. Such values are in the radio frequency range of the spectrum.

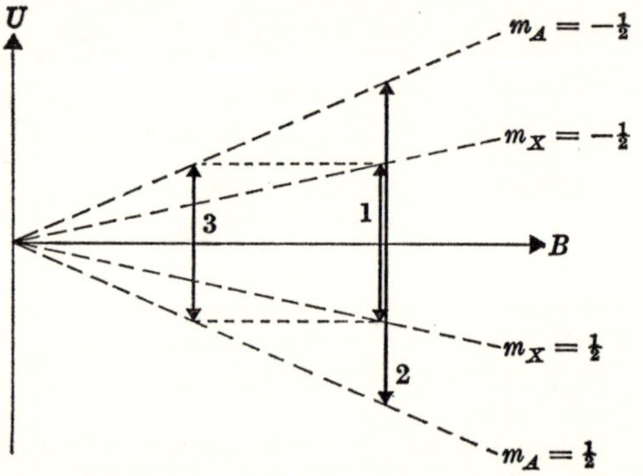

Fig. 1-3 Energy diagram for two different isolated nuclei of spin $I = \frac{1}{2}$, A and X, illustrating constant frequency and constant field conditions for resonance. The diagram is drawn with $\gamma_A > \gamma_X$ (both positive). If resonance conditions are such that transition 1 is stimulated for nucleus X, then to achieve resonance for A, either the frequency must be increased at constant field (transition 2), or the field must be decreased at constant frequency (transition 3).

Since the resonance frequency is proportional to magnetic field, either ν or B may be varied in order to achieve resonance. If it is required to change from observing the resonance of nucleus X to that of nucleus A, where the latter has a higher magnetogyric ratio than the former, then Eqn. (1-18) states that either ν may be *increased* or B may be *decreased* (see Fig. 1-3). Both methods have been used, but there are advantages in using the highest value of B consistent with magnetic field homogeneity requirements, so it is more normal to alter ν. For example, in a field of 2·35 T protons resonate at ca. 100 MHz whereas ^{11}B nuclei absorb at ca. 32 MHz. In contrast to the case just considered, where field and frequency have opposing effects as far as achieving resonance is concerned, for a given type of nucleus (i.e. γ constant), *increase* of B requires an *increase* of ν if resonance is to be maintained. Thus protons resonate at ca. 40 MHz in a 1 T field, and at ca. 60 MHz in a 1·5 T field.

1-6 Larmor precession

An alternative approach to NMR, which is frequently of great use, is to consider the classical motion of a magnetic moment in a uniform magnetic field under the condition of constant total energy. A torque is exerted on the magnetic moment which tends to align it perpendicular to the field, $\mathbf{B}_0$. However, since $\boldsymbol{\mu}$ moves, the direction of the torque is always changing, and the net result corresponds to a rotation of the direction of $\boldsymbol{\mu}$ about $\mathbf{B}_0$. The movement of $\boldsymbol{\mu}$ traces out a cone about $\mathbf{B}_0$. Such motion is referred to in general as precession (see Fig. 1-4) and in such instances as are considered here it is known as

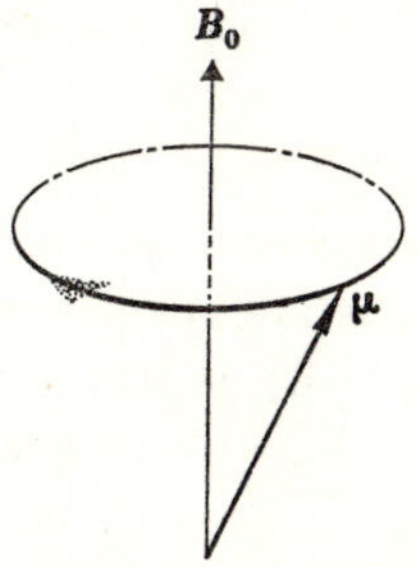

Fig. 1-4 Precession of a magnetic moment $\boldsymbol{\mu}$ about an applied magnetic field $\mathbf{B}_0$. It is assumed that the magnetogyric ratio is positive, as for the proton.

Larmor precession. The induced angular velocity of the precession is given by $\boldsymbol{\omega}_i = -\gamma \mathbf{B}_0$ (Section 4-2). Thus the Larmor frequency, ν_i, is

$$\nu_i = (\gamma/2\pi)\, B_0 \qquad (1\text{-}19)$$

Quantization of magnetic dipolar energy in the field B_0 can readily be incorporated in this picture; the half-angle of the cone of precession is determined by the quantum number m_I, i.e. the half-angle is given by $\cos\theta = m_I/[I(I+1)]^{1/2}$. Thus nuclei of spin I will be distributed among $2I + 1$ precession cones. The spin $\frac{1}{2}$ case is shown in Fig. 1-5. It should be noticed that the direction and frequency of precession are independent of m_I; when γ is positive the motion about $\mathbf{B}_0$ is left-handed as shown in Fig. 1-5.

Suppose there is an additional weak magnetic field, $\mathbf{B}_1$, perpendicular to $\mathbf{B}_0$. Such a field will also exert a torque on $\boldsymbol{\mu}$, tending to change the angle θ between $\boldsymbol{\mu}$ and $\mathbf{B}_0$. However, if $\mathbf{B}_1$ is fixed in direction it will alternately try to increase and decrease θ as $\boldsymbol{\mu}$ precesses. Since B_1 is stated to be weak, the net effect will be a slight wobbling in the precession of $\boldsymbol{\mu}$; such an effect is referred to as nutation. Alternatively, the motion of $\boldsymbol{\mu}$ can be described as caused by a resultant field $\mathbf{B}_0 + \mathbf{B}_1$. If, on the other hand, $\mathbf{B}_1$ is not fixed in direction,

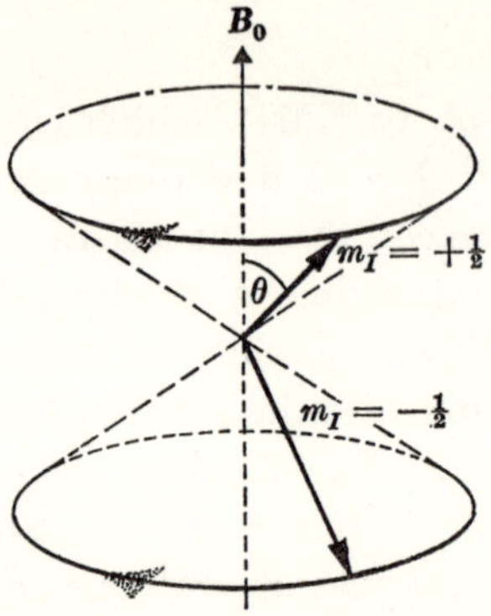

Fig. 1-5 The two precessional cones for a spin $\frac{1}{2}$ nucleus in a magnetic field, showing typical spin vectors. The angle θ is 54° 44′ in this case.

but is rotating about $\mathbf{B}_0$ with the same frequency as the precession of $\boldsymbol{\mu}$, its orientation with respect to $\boldsymbol{\mu}$ will be constant. Suppose this orientation is such that $\mathbf{B}_1$ is always perpendicular to the plane containing $\mathbf{B}_0$ and $\boldsymbol{\mu}$ as in Fig. 1-6; then the torque exerted on $\boldsymbol{\mu}$ by $\mathbf{B}_1$ will always be *away* from $\mathbf{B}_0$. Consequently, a large effect on $\boldsymbol{\mu}$ is possible (see Fig. 1-6 and Chapter 4). Since changing θ corresponds to changing the energy of $\boldsymbol{\mu}$ in $\mathbf{B}_0$, this condition is described as resonance—the frequency, ν, of the field $\mathbf{B}_1$ required must, by the above discussion, equal the Larmor precession frequency of Eqn. (1-19). This statement is clearly in agreement with the quantum-mechanical formula for NMR absorption given by Eqn. (1-18). The energy for the change of θ is, of course, derived from the rotating field B_1, which is supplied by radio frequency electromagnetic radiation.

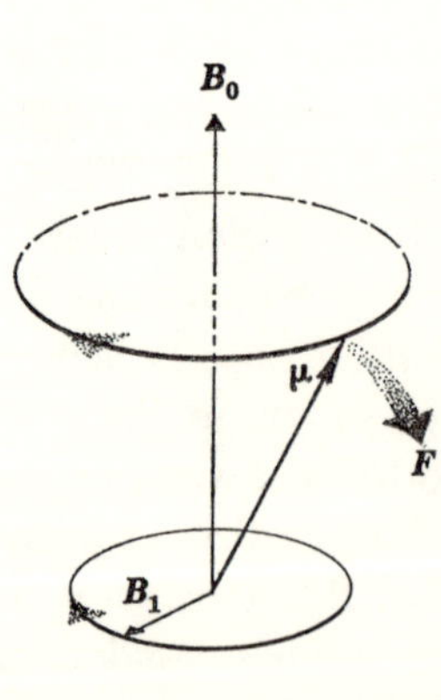

Fig. 1-6 The effect of a rotating magnetic field, $\mathbf{B}_1$, on a precessing magnetic moment, $\boldsymbol{\mu}$. When the B_0–B_1 plane is perpendicular to the B_0–μ plane there is a force **F** acting to increase the angle between $\mathbf{B}_0$ and μ. If $\boldsymbol{\mu}$ and $\mathbf{B}_1$ are rotating at the same rate, this force acts always away from $\mathbf{B}_0$ and therefore has a consistent effect.

1-7 Electronic shielding

The discussion of Sections 1-5 and 1-6 assumed that the presence of the sample does not perturb the magnetic field. However, this is not strictly true. Indeed, the nuclear Larmor precession itself produces a secondary magnetic field, but this effect is negligible compared to that produced by electrons. In fact, when a substance is placed in a magnetic field it becomes magnetized and modifies the field. The magnetization may be considered in terms of two contributions—a bulk (macroscopic) effect and a local (microscopic) effect. The bulk intensity of magnetization is proportional to the magnetic field, and the constant of proportionality is known as the bulk susceptibility, χ, of the material.* For substances with no unpaired electrons, χ is normally negative, and such substances are said to be diamagnetic:† the magnetization reduces the magnetic field within the sample. The bulk effect is, however, of little importance for NMR, so in later sections we shall be concerned only with the local effect.

The molecular interpretation of diamagnetism (both bulk and local effects) is straightforward: the behaviour may be traced to motion of electrons induced by the magnetic field. Figure 1-7(a) shows diagrammatically what happens to the electrons in an atom—they are induced to circulate around the nucleus about the direction of the applied field $\mathbf{B}_0$. It can be shown that the angular velocity is

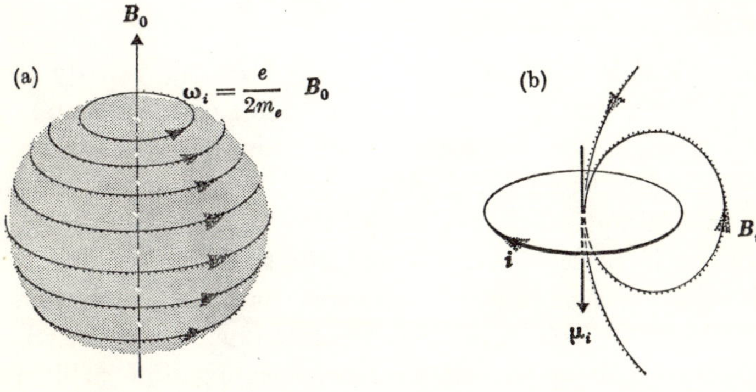

Fig. 1-7 Electronic shielding: (a) circulation of the electronic charge cloud under the influence of a magnetic field, (b) the secondary magnetic field produced by the precession. Note that the circulation is equivalent to electric currents in the reverse direction.

* Susceptibilities as normally listed are defined in terms of the magnetic field intensity H and not the magnetic induction field B (see the preface).

† Substances with unpaired electrons have permanent magnetic moments which cause χ to be positive (giving paramagnetism). Such compounds are important in electron spin resonance spectroscopy but are of only minor importance in NMR.

given by

$$\boldsymbol{\omega}_i = (e/2m_e)\mathbf{B}_0 \tag{1-20}$$

Note that the sign implies that the circulation is right-handed about $\mathbf{B}_0$ [Fig. 1-7(a)], i.e. in the opposite sense to proton spin precession (Fig. 1-4). Since the circulation involves motion of charge, there will be an associated induced magnetic moment, $\boldsymbol{\mu}_i$, which for a single electron at distance $\mathbf{r}$ from the nucleus is given by Eqn. (1-21) (see Problem 1-4)

$$\boldsymbol{\mu}_i = -(e^2/4m_e)r^2\mathbf{B}_0 \sin^2 \theta \tag{1-21}$$

where θ is the angle between $\mathbf{r}$ and $\mathbf{B}_0$. In order to find the total magnetic moment, it is of course necessary to sum over the whole electronic distribution. The induced magnetic field produced by $\boldsymbol{\mu}_i$ opposes the primary applied field $\mathbf{B}_0$ [Fig. 1-7(b)]. This is in fact essential if the law of conservation of energy is to hold (the situation is analogous to that in which Lenz's law is applicable). A negative magnetic susceptibility (i.e. diamagnetism) is implied. In principle, if enough information about electron densities and electronic wave functions were available, values of the bulk susceptibility could be calculated, since the intensity of magnetization can be equated to the total magnetic moment induced per unit volume.

It is clear that the electrons will shield (or screen) a nucleus from the influence of the field B_0. This shielding can be taken into account by using an effective field B at the nucleus, given by

$$B = B_0(1 - \sigma) \tag{1-22}$$

The dimensionless number σ is a small fraction, usually listed in parts per million, and is known as the shielding constant. In principle it may be calculated for atoms using equations similar to Eqn. (1-21). Therefore χ and σ are related quantities, but the relationship is not a simple one. A more detailed treatment of atomic screening will be given in Section 3-2. Clearly, the situation is more complex for molecules, because there is then more than one positively charged centre to influence the circulation of electrons. However, the same type of effect will occur. It is, moreover, possible arbitrarily to divide the total effect into contributions from the electrons associated with the various atoms and bonds in the molecule under consideration. Thus, we may speak of bond-induced magnetic moments and bond susceptibilities (see Section 3-7).

1-8 Chemical shifts

If the frequency of NMR transitions were entirely determined by Eqn. (1-18), the technique would have little chemical application.

Nuclei of different elements or isotopes would give resonances in different parts of the spectrum due to variations in γ, but this fact cannot be exploited easily for elemental analysis, because it is difficult to compare intensities of absorption at the very different frequencies involved. However, in 1949 it was found that the frequency of ^{14}N absorption depended on the chemical environment of the nucleus. This feature has since been shown to be general, and such differences in resonance conditions are referred to as chemical shifts. The existence of chemical shifts enables the NMR spectroscopist to distinguish between different chemical environments and makes NMR spectroscopy a powerful tool for the determination of the structure of molecules.

The phenomenon of the chemical shift arises because of shielding (or screening) of the nuclei from the external magnetic field by the electrons, discussed in the previous section. Since the shielding effect is caused by electronic environment, values of σ will vary with the position of the nucleus in the molecule. Factors that affect electron density will also affect σ. Thus the shielding constant for the proton of an aldehyde group is lower than that for the protons of a methyl group. The resonance condition [Eqn. (1-18)] becomes

$$\nu_j = |\gamma/2\pi|\, B_0(1 - \sigma_j) \tag{1-23}$$

The factors ν_j and σ_j refer to the resonance frequency (or Larmor frequency) and shielding constant respectively, for nucleus j. Variations in σ thus cause variations in resonance frequencies and are the reason for the occurrence of chemical shifts.

Strictly speaking, the appropriate field B_0 to use in Eqn. (1-20) and thereafter is not the applied field, but the applied field *corrected for the effect of bulk magnetization.* The correction depends on the sample shape and bulk susceptibility. Care must be taken in comparing shielding constants for nuclei in different compounds to either eliminate these bulk effects or make allowance for them. This is particularly important because, as discussed in the next section, absolute values of shielding constants are not normally obtained, but chemical shifts are reported relative to that for a standard substance.

For a given compound the appearance of the spectrum is governed by intramolecular chemical shift differences, i.e. differences in resonance frequencies for different nuclei of the same molecule. As an example, Fig. 1-8 shows 40 MHz and 60 MHz proton resonance spectra of chloroacetone: the CH_3 and CH_2 protons give absorption at different positions in the spectrum. Figure 1-8 illustrates another of the important features of NMR spectra, namely that the intensity

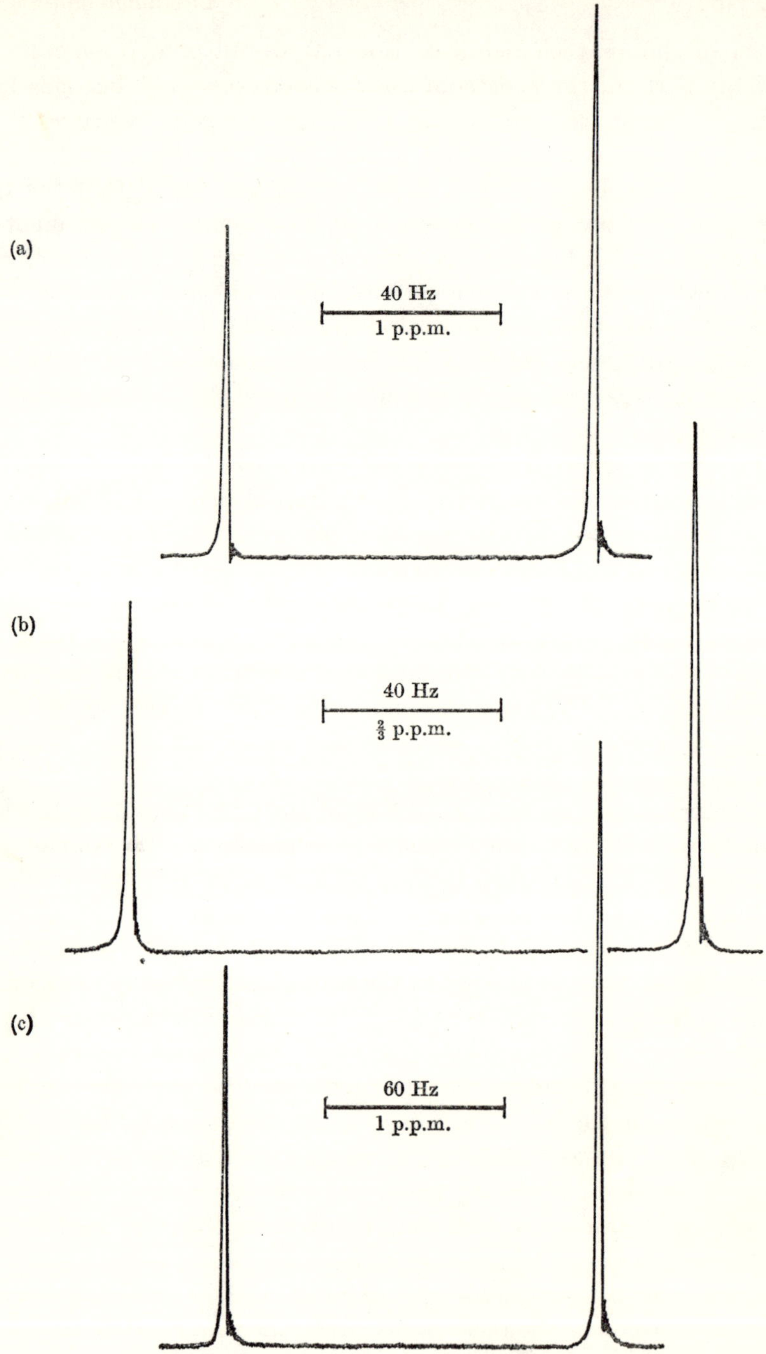

Fig. 1-8 PMR spectrum of chloroacetone: (a) at 40 MHz ($\equiv |\gamma B_0/2\pi|$), (b) at 60 MHz with the same scale as (a) in Hz, (c) at 60 MHz with the same scale as (a) in p.p.m.

of absorption is strictly proportional to the concentration of protons giving rise to the absorption. This is of great importance for structure determination. For instance, if the compound used in Fig. 1-8 was of unknown structure, the NMR spectrum would immediately show that it contained two types of proton and that the relative numbers of each type were 2:3. This use of NMR could be described as providing a relative proton count. Intensities in infrared and ultraviolet spectroscopy cannot be used in this way. Moreover, the proportionality of intensity to concentration for NMR is valid for comparisons between different molecules, thus providing a rapid non-destructive method of obtaining concentrations of impurities, isomer ratios etc. It is also a straightforward matter to obtain absolute concentrations for a given chemical in solution if standard concentrations can be prepared; this technique is of considerable application in quantitative analytical chemistry, for example for process control in chemical industry.

It may be seen from Eqn. (1-23) that resonance frequencies at constant field B_0 are directly proportional to the magnitude of B_0. Thus, all chemical shift differences (if measured in Hz) increase in proportion if B_0 is increased. It is therefore more usual to report chemical shift differences in parts per million (p.p.m.). These are obtained from measurements in Hz by dividing by ν_0, the operating frequency of the spectrometer ($\sim|\gamma B_0/2\pi|$). In such dimensionless units, chemical shift differences are invariant to the spectrometer conditions, and may therefore be used as molecular characteristics. Figures 1-8(a) and 1-8(b) show how an increase in B_0 increases chemical shift differences in Hz, while Fig. 1-8(c) shows the same spectrum as in Fig. 1-8(b), but with calibration in p.p.m. on the same scale as Fig. 1-8(a).

1-9 Reference standards and solvents

Ideally, the chemist would like to know the absolute magnitudes of shielding constants. This would require an accurate knowledge of both B_0 and the resonance frequency. However, it is not normally feasible to measure magnetic fields to a high accuracy; on the other hand, it is usually convenient and accurate to measure small differences in resonance frequency under constant field conditions. Consequently, it has become normal practice to measure resonance frequencies relative to that for certain nuclei in a standard chemical. Chemical shift values are thus quoted relative to such a standard. The standard chemical may be either external or internal to the sample being studied. When used externally, it is placed in a capillary tube surrounded by the sample. However, in such cases the magnetic field B_0 for the sample is not strictly comparable with that for the standard, because the liquids will have different magnetic susceptibilities. Consequently bulk-susceptibility corrections need to be made before chemical shift differences can be quoted. This procedure is to be avoided where possible, and therefore we shall discuss internal references only. However, measurements made with an external standard in cases of known chemical shifts may be used in an NMR method of obtaining magnetic susceptibility.

Measurements with internal standards are made by adding a small amount (ca. 5%) of the standard directly to the sample under study. Susceptibility effects are then the same at both standard and sample molecules (i.e. the value of B_0 is the same), and observed chemical shift differences can be attributed to local influences only. Clearly, a suitable standard should possess as many of the following characteristics as possible:

(a) It should contain only a single type of magnetic nucleus with a single type of environment, so that the NMR spectrum consists of one sharp line.

(b) It should contain many nuclei of that type and environment, so that an intense signal is given, even when the substance is present only in small amount.

(c) The resonance should occur in a region of the spectrum away from the signals given by most common chemicals.

(d) The standard should be chemically inert, so that its resonance position is not affected by the sample or vice versa. For this criterion to apply, there must be no transient complex formation with solvent or solute molecules, and no irreversible chemical reaction. Since most solvent effects are due to electrostatic or steric interactions, the standard chemical should therefore be non-polar and highly symmetrical.

The substance which has become the accepted standard for proton resonance is tetramethylsilane, $Si(CH_3)_4$, which will be referred to as TMS throughout the rest of this book. The most common isotopes of silicon and carbon have zero spin, and the proton resonance of TMS occurs to high field of most other proton resonances. Due to rapid internal rotation about the Si–C bonds, all the protons are equivalent. Thus all four criteria are satisfied; TMS has the added advantage that it is a volatile liquid that can be readily handled (though its volatility is for some purposes a little higher than is desirable). Other reference standards are normally necessary for resonances of nuclei other than protons. As with intramolecular chemical shifts, those relative to a standard are proportional to the spectrometer operating frequency if measured in Hz; consequently values are reported in the dimensionless p.p.m. units. Proton resonance chemical shifts in p.p.m. relative to TMS are said to be based on the δ scale. The δ values are positive if the sample absorbs to high frequency of the reference absorption at constant field, as expressed by Eqn. (1-24):*

$$\delta = 10^6(\nu_{\text{sample}} - \nu_{\text{TMS}})/\nu_{\text{TMS}} \text{ at constant } B_0 \qquad (1\text{-}24)$$

Since shielding constants are small, the value of ν_{TMS} in the denominator may be replaced by the spectrometer operating frequency ν_0 or by $|\gamma B_0/2\pi|$ with little error. Thus δ is related to shielding constants by Eqn. (1-25):

$$\delta = 10^6(\sigma_{\text{TMS}} - \sigma_{\text{sample}}) \qquad (1\text{-}25)$$

It is clear that a high frequency shift implies a decrease in σ, i.e. deshielding. Figure 1-9 shows the spectrum of a mixture of benzene and acetone, including internal TMS. The δ scale is shown, together with the calibration in Hz at 60 MHz.

Occasionally it becomes necessary to employ secondary standards and to obtain the chemical shifts of the sample indirectly. This is required for work at high temperatures (when TMS vaporizes) or when TMS is insufficiently soluble, as occurs for certain highly polar solvents. The case most commonly met with is aqueous solutions. The salt sodium 2,2-dimethyl-2-silapentane-5-sulphonate, $Me_3SiCH_2CH_2CH_2SO_3Na$, has been suggested for such work (although not all the criteria given above are satisfied) because the resonance of the methyl protons occurs so close to $\delta = 0{\cdot}00$ that the difference can usually be neglected.

As has been implied above, most high-resolution NMR work is done with samples in the liquid phase. The reasons for this will be developed in Chapter 4, but at this stage it is desirable to consider suitable solvents, since these will be necessary for solid samples. Clearly, solvents which do not contain the isotope whose resonance is being studied are the most suitable. For proton resonance CCl_4 is very widely used. However, it is not useful for polar solutes and it is often convenient to use deuterated solvents; the most common ones (to a considerable extent because they are relatively cheap) are D_2O and $CDCl_3$. It is sometimes necessary to use fully deuterated benzene, acetone, or dimethylsulphoxide. In cases where the solvent resonances do not obscure those of the sample, it is not worth using

* An alternative to the δ scale is that known as the τ scale. This lists chemical shifts in p.p.m. relative to TMS but with increasing values to low frequency such that the resonance of TMS occurs at $\tau = 10{\cdot}00$. The relationship between the τ and δ scales is given by $\tau = 10{\cdot}00 - \delta$. The chemical shift scales used for nuclei other than protons are also often defined as positive to low frequency, in contrast to the δ scale.

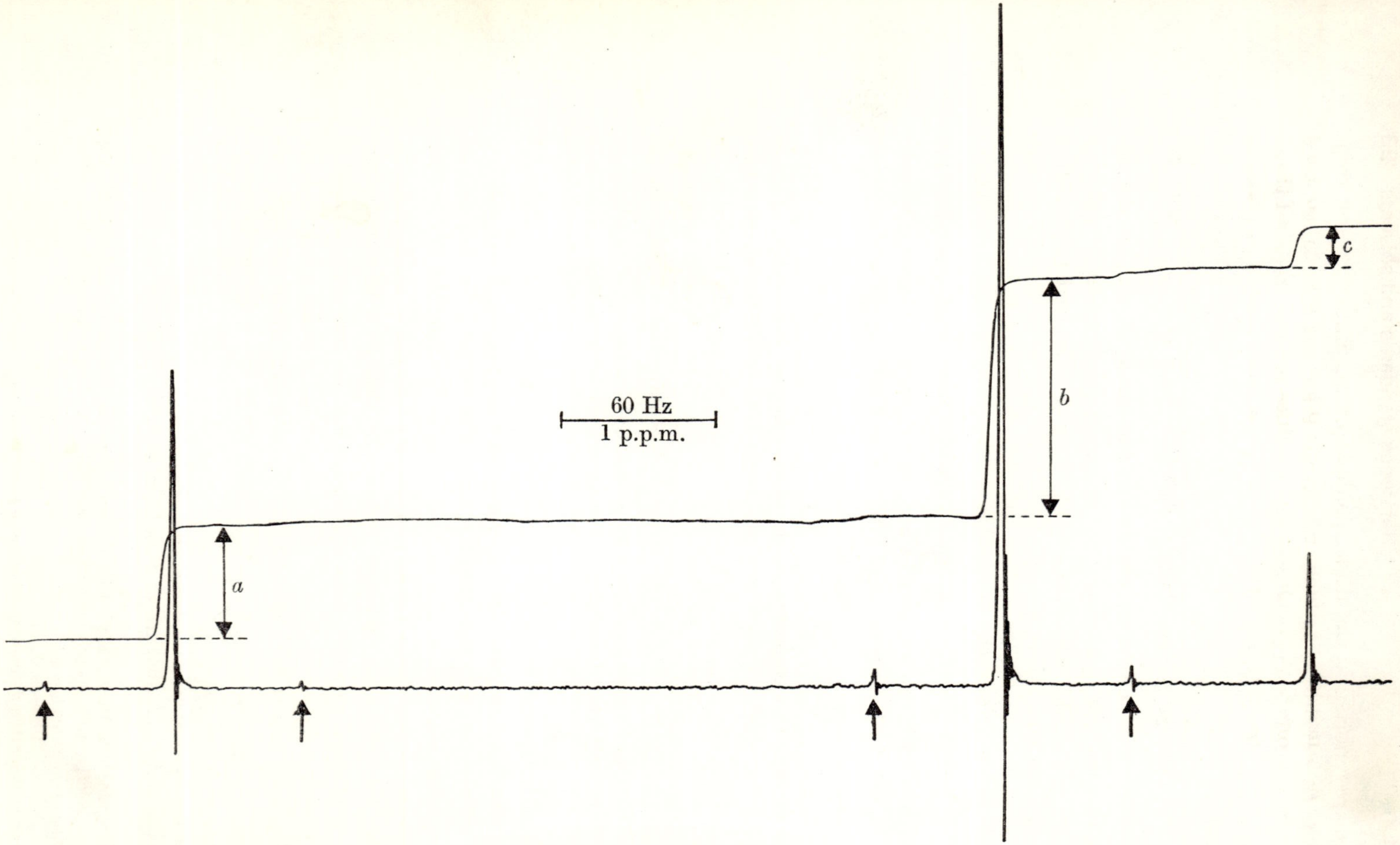

Fig. 1-9 PMR spectrum (at 60 MHz) of a mixture of benzene and acetone with TMS (the three resonances are in that order from left to right). The resonances of benzene and acetone occur at δ 7·33 and δ 1·97 in this solution. The components of the solution are approximately in the molar ratio 10:22:1, as can be verified from the integral trace superimposed on the spectrum (see Section 1-17)—the steps a, b, and c are proportional to concentration in terms of numbers of protons per unit volume. The small peaks marked with arrows are spinning side bands (see Section 1-17); the spinning rate is ca. 50 Hz.

deuterated solvents. The most suitable chemicals are those with only one chemical type of proton (and no other nuclei of spin $I = \frac{1}{2}$) since these obscure only a small region of the spectrum. The δ values of several suitable solvents are given in Table 1-2. The dielectric constants listed are helpful in deciding the likelihood of a given solute being soluble. A further feature that governs the choice of solvent is the temperature at which the spectrum is to be taken.

Table 1-2 Chemical shifts, melting and boiling points, and dielectric constants of some useful NMR solvents

	$\delta^{(a)}$	*Melting point* (°C)	*Boiling point* (°C)	$\kappa^{(b)}$
Trifluoroacetic acid	11·61	−15	72	39·5
Benzene	7·37	6	80	2·3
Chloroform	7·27	−64	61	4·8
Trichloroethylene	6·45	−73	87	3·4 (ca. 16°C)
1,1,2,2-Tetrachloroethane	5·95	−36	146	8·2
Methylene dichloride	5·30	−97	41	9·1
Water	4·76	0	100	80·4
p-Dioxane	3·70	10	101	2·2 (25°C)
Dimethylsulphoxide	2·62	6	$100^{(c)}$	46.0 (ca. 35°C)
Acetone	2·17	−95	57	20·6
Acetonitrile	2·00	−42	82	37·5
Cyclohexane	1·43	7	81	2·0

(a) All δ values except for water and trifluoroacetic acid are for ca. 7% solutions in $CDCl_3$ (plus TMS).
(b) Dielectric constants at ca. 20°C except where stated.
(c) Unstable at high temperatures.

Thus the melting and boiling points of the solvent must be taken into consideration (see Table 1-2). For low-temperature work, CS_2 (melting point −109°C) often takes the place of CCl_4 (melting point −23°C); acetone and toluene (the latter in spite of the fact that it contains more than one type of proton) are also useful low-temperature solvents. Highly chlorinated compounds such as tetrachloroethylene are frequently employed for high-temperature work.

1-10 Spin–spin coupling

In most spectra additional features are present. In the simplest cases (referred to as first-order spectra) these occur as a splitting of the bands due to each nucleus. If there are only two nuclei with non-zero spin in the molecule under consideration, having, say, spins

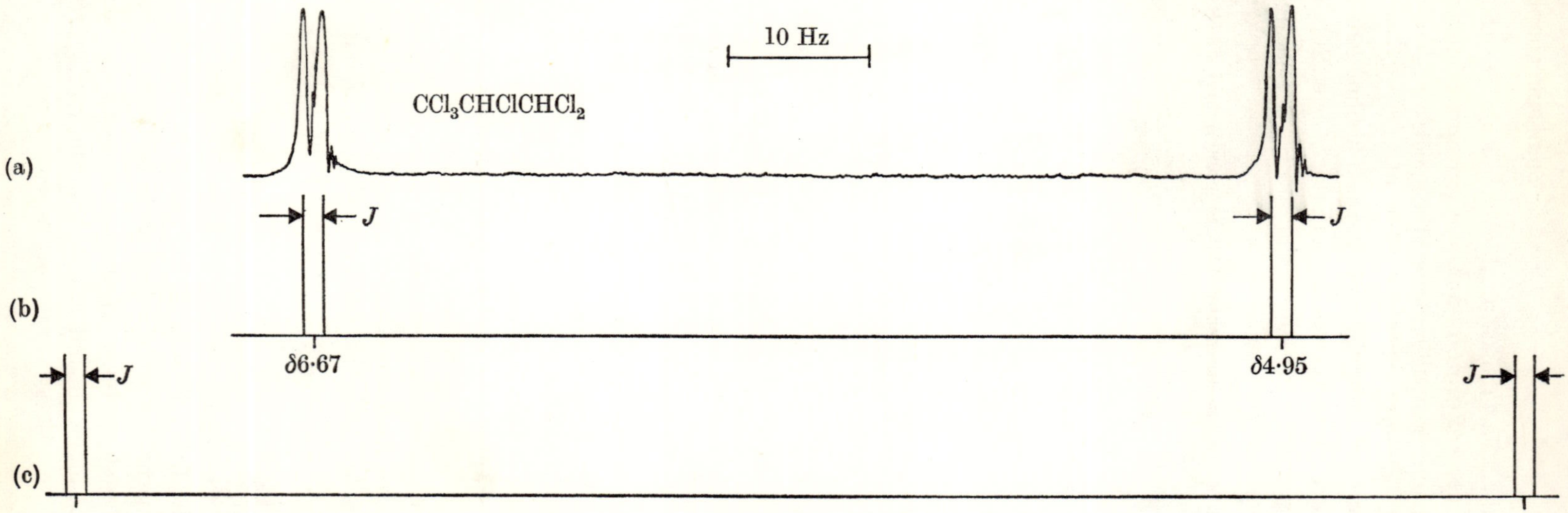

Fig. 1-10 A first-order spectrum of a two-spin system: the proton resonances of 1,1,1,2,3,3-hexachloropropane (a) observed at 40 MHz, (b) first-order lines corresponding to (a), (c) first-order lines calculated for 60 MHz (with the average chemical shift held at the same position). The coupling constant is 1·4 Hz and the chemical shift difference is 1·72 p.p.m. The pair of lines at high frequency is attributed to the —$CHCl_2$ proton. [We are grateful to A. B. Dempster and N. Sheppard for spectrum (a).]

I_1 and I_2, then it is found that the resonance of spin I_1 is split into $2I_2 + 1$ lines of equal intensity and that of spin I_2 is similarly split into $2I_1 + 1$ lines. The line separations are all equal. The nature of the splitting suggests, correctly, that the different lines of I_1 arise from molecules with differing values of m_I for the second spin. In fact the energies of I_1 are modified by a small interaction between the spins. This interaction is known as spin–spin coupling. Its magnitude is determined by a spin–spin coupling constant (usually simply referred to as a coupling constant, and written J_{jk} for interaction between spins j and k). The transition energies are affected by coupling, and indeed the splittings in the spectra are equal in magnitude to J. When both nuclei have spin $I = \frac{1}{2}$, the spectra consist of two doublets, i.e. one doublet assigned to transitions of each spin. The doublet splittings are equal to J. The example in Fig. 1-10(a) is the proton spectrum at 40 MHz of $CCl_3CHClCHCl_2$.

Coupling constants, like chemical shifts, depend on chemical environment and are therefore of great use in structure determination. However, variation of the spectrometer operating frequency does *not* affect the magnitude of the spectral splittings due to coupling, expressed in Hz (contrast the case of chemical shift differences). Therefore coupling constants are always expressed in Hz and never in p.p.m. Because chemical shifts and coupling constants are affected differently by changes in spectrometer conditions, the spectra themselves change in appearance. Figures 1-10(b) and (c) show theoretical proton spectra at 40 MHz and at 60 MHz to illustrate this point. Individual absorption lines should never be assigned δ values, since their positions are unlikely to be independent of spectrometer operating frequency; the δ scale should be reserved for true chemical shift positions, or simply for a general indication of a region of the spectrum.

When there are more than two magnetic nuclei in the molecule, coupling may occur between each pair of nuclei, but the coupling constants will normally all be different in magnitude unless some nuclei are related by symmetry. Each coupling constant J_{jk} causes splittings in the resonance of nuclei j and k (but not that of other nuclei). The pattern of lines for a given nucleus may be explained by the method of successive splittings. Consider a system consisting of three spin $\frac{1}{2}$ nuclei, which will be labelled A, M, and X. The A resonance is split by coupling to both M and X, but to different extents. Let us assume $J_{AM} > J_{AX}$. The pattern of A lines may be predicted by considering first of all a splitting by M, and then a further splitting of each of the resulting pair of lines by X. The final pattern is of four lines. This procedure is illustrated in Fig. 1-11.

Note that J_{AM} is given by the separation of lines 1 and 3 or 2 and 4 while J_{AX} may be measured from the splitting between lines 1 and 2 or 3 and 4. The M resonance will be split similarly but not identically (since this time J_{AM} and J_{MX} are involved). Altogether the spectrum consists of 12 lines (4A, 4M, and 4X). It is important to realize that each part of the spectrum is symmetrical about its mid-point, and

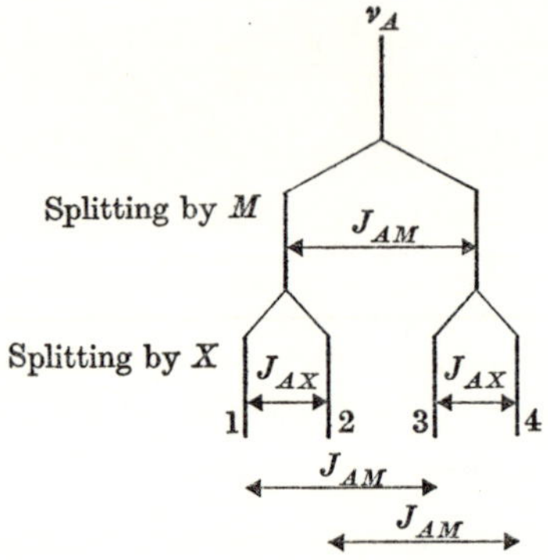

Fig. 1-11 The method of successive splittings for nucleus A of a three-spin system, showing the effect in first order of coupling to nuclei M and X.

that the mid-point corresponds to the Larmor frequency (or unperturbed resonance frequency) of the nucleus concerned. The full spectrum of a three-spin proton system is shown in Fig. 1-12.

The above discussion may be extended to show that for a system consisting of n spin $\frac{1}{2}$ nuclei (none of which are equivalent), each nucleus gives rise in principle to 2^{n-1} lines. The total number of lines is therefore $n2^{n-1}$. However, the possibilities that some lines may overlap, or that some coupling constants may be negligibly small, should be borne in mind. In fact, as a general rule the magnitude of coupling constants tends to decrease as the number of chemical bonds separating the coupled nuclei increases. This is because the most common mechanism for coupling involves the transmission of the effect through the bonding electrons; the magnitude of the affect is attenuated (when the electrons are localized into two-centre bonds) by the requirement that the spin information must be passed through successive bonds. For saturated hydrocarbons the coupling constant between protons separated by more than three bonds is usually negligibly small. Thus, spectral splittings may often be attributed to nearest neighbours only. However, such a generalization must be treated with great reserve, and frequently does not hold when there is unsaturation in the molecule.

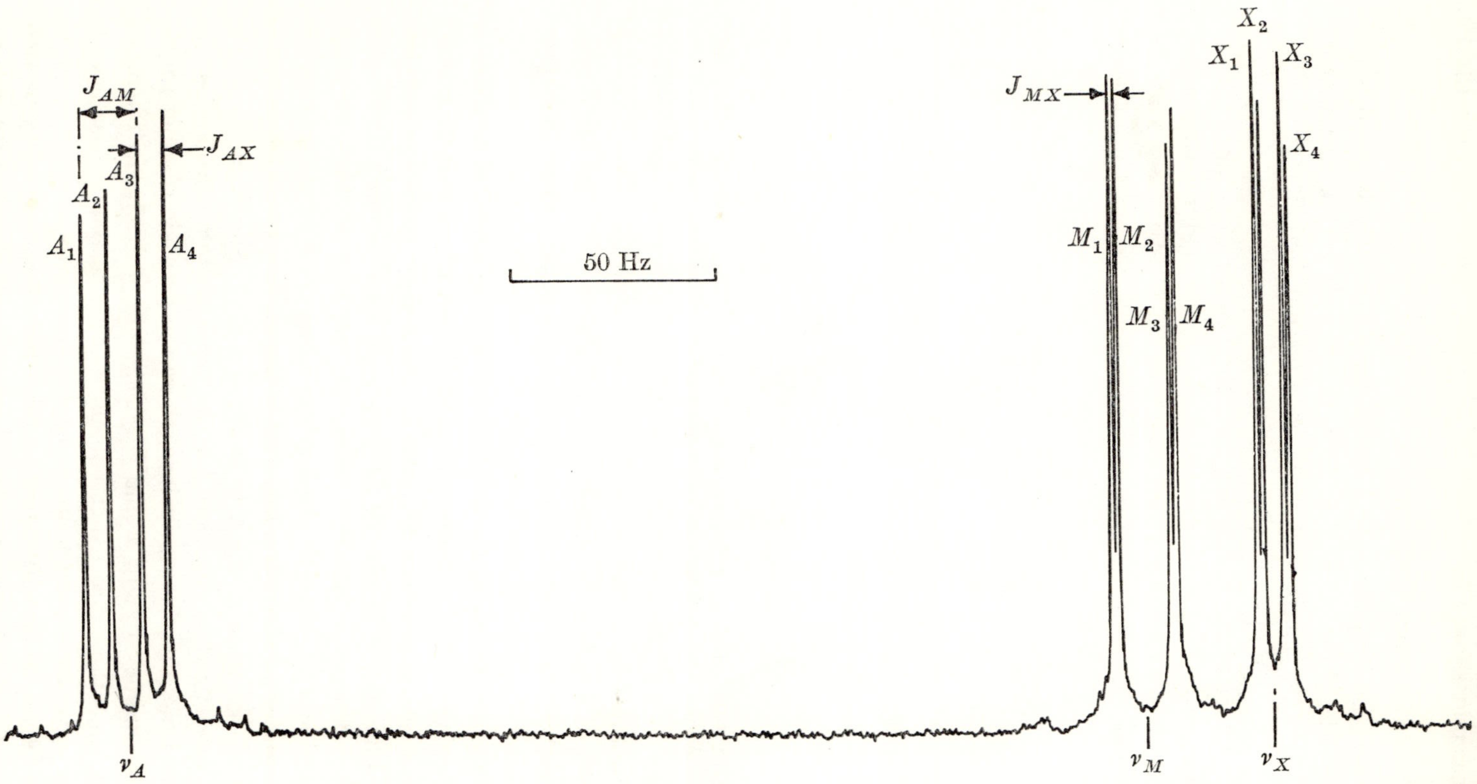

Fig. 1-12 100 MHz proton resonance spectrum of the vinyl protons in vinyl acetate (10% solution in CCl_4). The methyl protons are sufficiently distant in the molecule not to couple appreciably to the vinyl protons, which therefore form an isolated three-spin system. The spectrum is nearly first-order; the slight variations in intensity are due to second-order effects. The lines are labelled as in the energy level diagram of Fig. 1-14 under the assumption that all the coupling constants are positive, though for this compound the *geminal* coupling constant is known to be negative [see Problem 1-8(b)].

1-11 Basic product functions

The wave functions for spin $\frac{1}{2}$ nuclei are frequently labelled α for $m_I = +\frac{1}{2}$ and β for $m_I = -\frac{1}{2}$. This notation is very convenient in use and has the merit that it emphasizes our inability to specify the wave function more mathematically (the values of m_I are simply the eigenvalues of the particular operator $\hat{I}_z$—see Chapter 2).

In spectra such as those discussed above, where all the theoretical splittings due to coupling are resolved, each transition may be labelled with the spin states (α or β values) of the nuclei other than

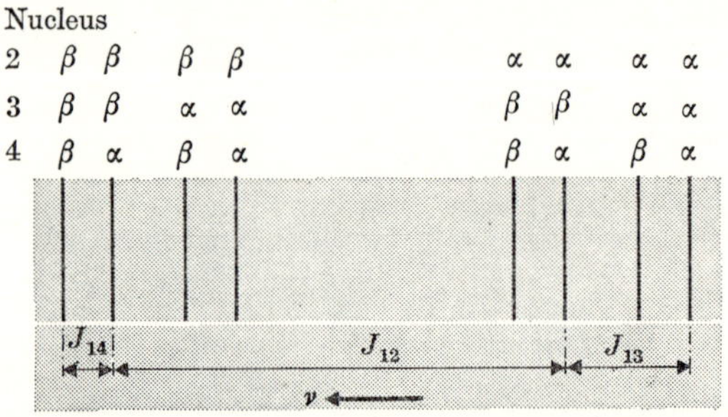

Fig. 1-13 The labelling of the transitions of nucleus 1 of a four-spin first-order system, according to the spin wave functions of the other three nuclei.

the one undergoing the transition. Thus the eight lines due to one of the nuclei of a four-spin system may be labelled as in Fig. 1-13 (the assignment actually assumes all the coupling constants are positive, which is not necessarily the case).

The total spin states for a system of spin $\frac{1}{2}$ nuclei may also be labelled using the α, β notation. In accordance with normal quantum-mechanical considerations the overall state may be designated by the product of the functions for the individual nuclei. Thus one state of a three-spin system (nuclei A, M, and X, say) may be designated $\alpha_A\beta_M\beta_X$. If the product functions are always listed using the same order for the individual nuclei, the subscripts can be omitted, and the above state referred to simply as $\alpha\beta\beta$. Such functions are termed 'basic product functions.' Since there are two possibilities for the wave function of each nucleus, there are 2^n basic product functions for a spin system of n spin $\frac{1}{2}$ nuclei. The word 'basic' refers to the fact that the more exact calculations discussed in Chapter 2 frequently use the product functions as a *basis set*.

It is useful to classify basic product functions according to the value of the total component of spin angular momentum, i.e. according to $m_T = \sum_j m_j$, where the summation is over all the nuclei

Table 1-3 The basic product functions of a three-spin system, with classification according to the total component of spin

m_T	*Basic product functions*
$-\frac{3}{2}$	$\beta\beta\beta$
$-\frac{1}{2}$	$\beta\beta\alpha$, $\beta\alpha\beta$, $\alpha\beta\beta$
$+\frac{1}{2}$	$\beta\alpha\alpha$, $\alpha\beta\alpha$, $\alpha\alpha\beta$
$+\frac{3}{2}$	$\alpha\alpha\alpha$

of the system. Table 1-3 lists the basic products with their m_T values for a three-spin system.

1-12 First-order energies and spectra

The simple spectra described in the preceding sections only arise when all chemical shift differences $|v_j - v_k|$ are much greater than the corresponding coupling constants $|J_{jk}|$. In such cases the basic product functions are eigenfunctions of the spin Hamiltonian, and the spectra are said to be first order. Each spin $\frac{1}{2}$ nucleus gives rise to a set of resonances well-separated from those of the other nuclei. The group of lines due to a given nucleus is symmetrical about its chemical shift position and the splittings within the group are equal to the appropriate coupling constants. In this section some of the basic features of first-order systems will be discussed. Situations where second-order effects are important will be treated in Chapter 2.

In Section 1-6 it was stated that the selection rule for NMR transitions of a single nucleus of spin I is $\Delta m_I = \pm 1$. For a spin $\frac{1}{2}$ nucleus this means that the transition $\alpha \rightarrow \beta$ is allowed in absorption. For a spin system consisting of several spin $\frac{1}{2}$ nuclei the selection rule $\Delta m_T = \pm 1$ is rigidly obeyed. Moreover, an allowed transition involves a change in the spin component of only one nucleus. Thus the selection rule for a transition assigned to nucleus j is that of Eqn. (1-26).

$$\Delta m_j = \pm 1; \qquad \Delta m_k(k \neq j) = 0 \tag{1-26}$$

Thus the simultaneous change of three spins (e.g. $\alpha\alpha\beta \rightarrow \beta\beta\alpha$ for a three-spin system) is forbidden. With these considerations in mind the energy level diagram (showing allowed transitions) for a system of three spin $\frac{1}{2}$ nuclei is as shown in Fig. 1-14. This energy level diagram satisfactorily indicates the number and type of transitions discussed in Section 1-10.

For the detailed evaluation of energy levels, however, it is necessary to consider the energy equation. The expression appropriate to the first-order case is given as

$$h^{-1}U = -\sum_j v_j m_j + \sum_{j<k} J_{jk} m_j m_k \tag{1-27}$$

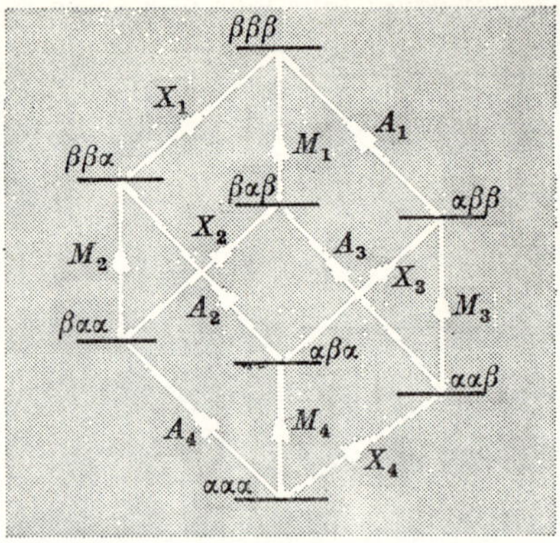

Fig. 1-14 Energy level diagram for a system of three spin $\frac{1}{2}$ nuclei, with the assignment of the allowed NMR transitions. The relative positions of the levels of the same value of m_T are only diagrammatic. The transition labels are the same as designated in Fig. 1-12.

where $\nu_j = (\gamma_j/2\pi)\ B_0(1 - \sigma_j)$, as in Eqn. (1-23), and is the Larmor frequency of nucleus j. The first term is the chemical shift term (often referred to as the nuclear Zeeman energy) and takes account of the interactions of the nuclear magnetic moments with the magnetic field. The second term is the coupling constant contribution and is due to electron-coupled interactions between the nuclear magnetic moments. The condition $j < k$ is written into the summation to ensure that a given pair of nuclei is included only once.

The way in which the energy level diagram is based on Eqn. (1-27) may be best discussed using Fig. 1-15, which shows the one- and two-spin cases. For the two-spin case there are four basic product functions, $\alpha\alpha$, $\alpha\beta$, $\beta\alpha$, and $\beta\beta$; on labelling the nuclei A and X Eqn. (1-27) becomes

$$h^{-1}U = -\nu_A m_A - \nu_X m_X + J_{AX} m_A m_X \tag{1-28}$$

Figure 1-15 shows the energy (a) for nucleus A alone, in the absence of the field **B**, (b) for nucleus A alone, in the presence of **B**, (c) for both nuclei A and X in the field, but with $J_{AX} = 0$ (i.e. using the chemical shift term only), and (d) for both A and X in the field with $J_{AX} \neq 0$. With selection rule (1-26) the allowed transitions are:

A transitions $\Delta m_A = \pm 1\ \Delta m_X = 0$

X transitions $\Delta m_A = 0\ \Delta m_X = \pm 1$

These are indicated on Fig. 1-15, which shows how coupling splits the two A transitions, which were degenerate for $J_{AX} = 0$, such that the splitting is $|J_{AX}|$. The spectrum is indicated in Fig. (1-16). It can be seen, both from Fig. 1-15 and from Eqns. (1-26) and (1-27)

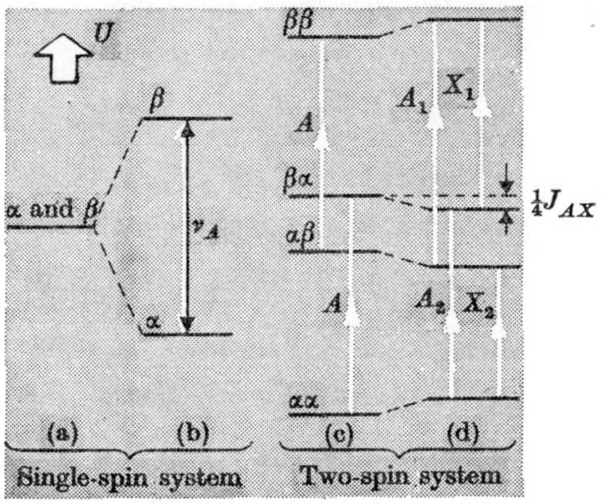

Fig. 1-15 Construction of the energy level diagram for one and two-spin systems on the basis of Eqn. (1-27). For further explanation see the text.

that the effect of the coupling term is to lower the energy by $\frac{1}{4}J$ if the coupled spins are 'anti-parallel' (i.e. $\alpha\beta$ or $\beta\alpha$) but to increase the energy by $\frac{1}{4}J$ if they are 'parallel' ($\alpha\alpha$ or $\beta\beta$). Actually, this is so only if J is positive (see Chapter 3). In fact Eqn. (1-28) defines the sign of a coupling constant as positive if coupling stabilizes anti-parallel spins and as negative if coupling destabilizes anti-parallel spins. This discussion is readily expanded to larger spin systems. The spectra predicted are those described in preceding sections. Actually, Eqn. (1-27) is applicable to nuclei of any spin, although we have applied it to the spin $\frac{1}{2}$ case.

One feature of Eqn. (1-27) and the resulting spectra that has not yet been noted is that although the energy levels depend on the signs of the coupling constants, the transition energies do not. Thus first-order spectra are invariant to a change of sign of any or all of the coupling constants. Neither relative nor absolute signs of coupling constants can be obtained from such spectra unless additional techniques, such as double resonance, are employed. Thus, strictly

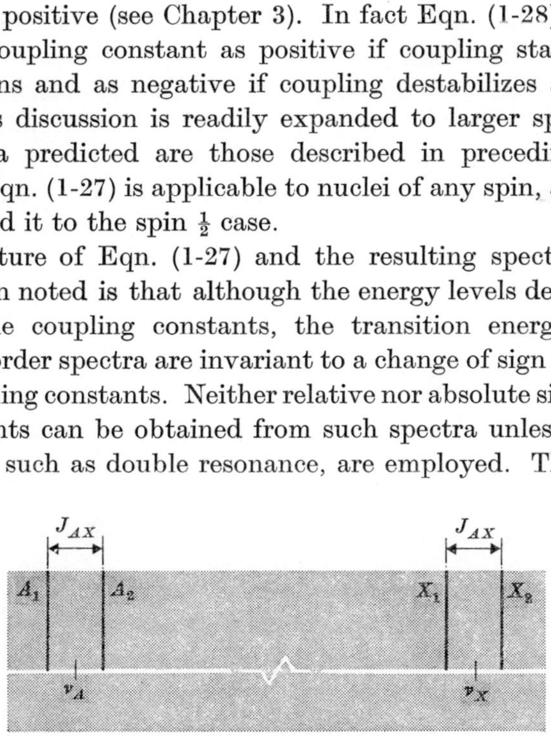

Fig. 1-16 The first-order NMR spectrum of a two-spin system (diagrammatic), derived from the energy level diagram. For the transition numbering see Fig. 1-15. For comparison with an observed spectrum see Fig. 1-10. Note agreement with the method of successive splittings.

speaking, first-order splittings are equal to the *moduli* of the appropriate coupling constants.

1-13 Equivalence

So far we have dealt with nuclei which differ in their electronic environment. Frequently, however, due to molecular symmetry, two or more nuclei may have identical environments. In such cases the nuclei (or atoms) are said to be equivalent. All physical and chemical properties will then be the same for the equivalent nuclei; for example, the rates of any chemical reaction will be the same at equivalent sites. Consequently it can be seen that symmetrically equivalent nuclei will have the same chemical shift (since this depends solely on environment). Nuclei with the same chemical shift are said to be isochronous. However, there are several ways in which nuclei can be equivalent, and it is important for NMR to distinguish two such cases. Consider the way in which two isochronous nuclei are coupled to a third magnetic nucleus which is not equivalent to them. The strength of the coupling may be equal for the two isochronous nuclei or it may be unequal. In the former case (if this holds for coupling to every individual nucleus that is non-equivalent to the two considered) the isochronous nuclei are *magnetically* equivalent, in the latter case they are *chemically* equivalent. Some examples will make this clear. In 1-fluoro-3,4,5-trichlorobenzene [**1**] the only spin $\frac{1}{2}$ nuclei are fluorine and hydrogen. The two protons are isochronous, by symmetry. They are also equally coupled to the fluorine nucleus, again due to symmetry (the relationship is an *ortho* one in the case of each (H, F) pair). Thus the protons are magnetically equivalent.

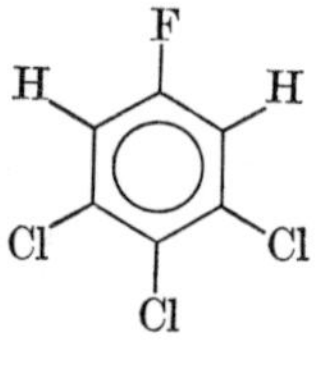

[**1**]

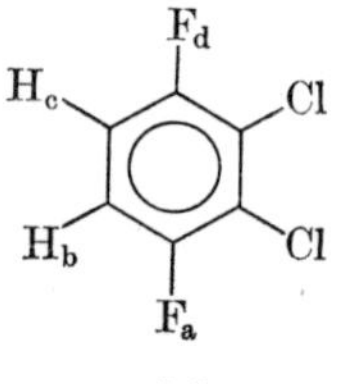

[**2**]

In the case of 1,4-difluoro-2,3-dichlorobenzene [**2**] the two protons are again isochronous (as are the two fluorine nuclei). However, if F_d is chosen as the 'third' nucleus, it can be seen that the (H_b, F_d) relationship is *not* the same as the (H_c, F_d) one—they are *meta* and *ortho* respectively. The corresponding coupling constants are therefore unequal, and the protons are magnetically *non*-equivalent; they are merely *chemically* equivalent. It is perfectly possible to have a situation with two isochronous pairs of nuclei which do have magnetic equivalence. An example is the complex [**3**], in which the two *trans* fluorine nuclei are magnetically equivalent because they are equally coupled to any one of the remaining fluorines; similarly the two remaining fluorine nuclei form a magnetically equivalent pair.

There are two types of situation which require further explanation. In the first place, if *all* the magnetic nuclei in the molecule are isochronous, they are necessarily magnetically equivalent, since there is no other nucleus to allow distinction between chemical and magnetic equivalence. Benzene, methyl iodide and TMS (provided rotation about the C–Si bonds is assumed to occur rapidly) are examples of this situation. Secondly, the more complicated cases [**4**] and [**5**] will be used to indicate an extension of the above

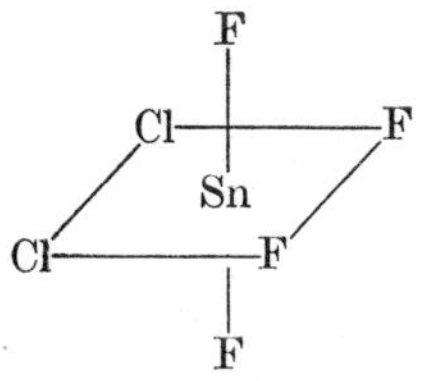

[**3**]

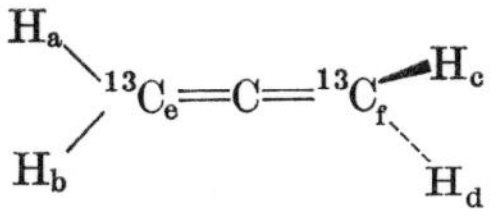

[**4**]

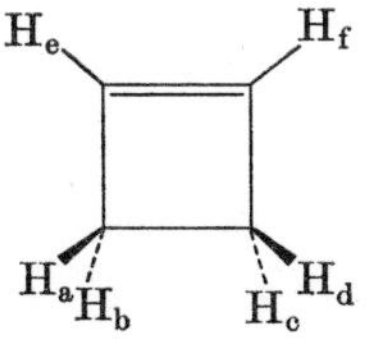

[**5**]

principles. It is clear that in both cases the four protons a, b, c, and d are all isochronous (symmetrically equivalent); on the other hand, the pairs a, b are not magnetically equivalent to the pairs c, d, since the two pairs couple differently to nucleus e. It is, however, more difficult to decide whether protons a and b are magnetically equivalent or not. The situation is resolved by considering how a and b couple to c, which is *not* part of the set a, b, since it is not magnetically equivalent to them. In [**4**], $J_{ac} = J_{bc}$ by symmetry (and $J_{ad} = J_{bd}$), so that a and b are magnetically equivalent (as are c and d). For [**5**], on the other hand, $J_{ac} \neq J_{bc}$ and $J_{ad} \neq J_{bd}$; therefore a and b are not magnetically equivalent, but are only chemically equivalent. However, it is not necessary for all the nuclei of a group to be equally coupled together for them to be magnetically equivalent. For instance, pentafluorosulphur chloride [**6**] has four isochronous fluorine nuclei in equatorial positions, a, b, c, and d, plus a lone axial fluorine,

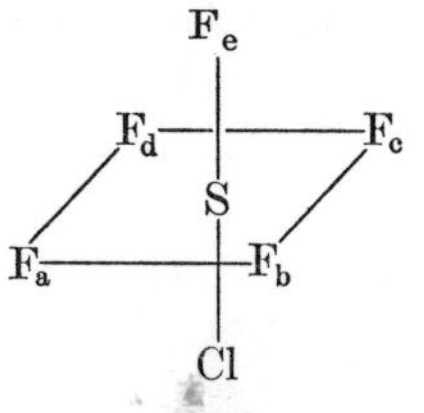

[**6**]

e. All the equatorial fluorine nuclei are equally coupled to the axial nucleus; a, b, c, and d are therefore magnetically equivalent. However, these nuclei are not equally coupled to one another, since $J_{ac} = J_{bd}$ (*trans* coupling) $\neq J_{ab} = J_{bc} = J_{cd} = J_{da}$ (*cis* coupling).

In general, both chemical and magnetic equivalence give rise to simplifications in the spectra; however, the simplifications are much greater for magnetic than for chemical equivalence. Indeed, except when there are certain fortuitous relationships between coupling constants, spectra from molecules containing chemical equivalence can never be first order (See Chapter 2). In certain situations, molecules with magnetically equivalent nuclei *can* give rise to first-order spectra. Magnetic equivalence in such circumstances is further considered in Section 1-14 when the appropriate notation has been developed; the more complex effects of magnetic equivalence will be discussed in Section 2-11.

1-14 Notation for spin systems

It is important to classify molecules or molecular fragments according to their spin systems, which define the number and type of magnetic nuclei and the relationship between them. The notation which will be described is normally used for the spin $\frac{1}{2}$ nuclei of a molecule only. Each nucleus is assigned a capital letter of the Roman alphabet. If the chemical shift difference between a pair of nuclei is much greater than the coupling constant between them, they are assigned letters well apart in the alphabet. This always applies if the nuclei are of different isotopes or elements. Thus *cis*-1,2-dichlorofluoroethylene [**7**] is said to possess an AX spin system (it is irrelevant whether A refers to H and X to F or vice versa). If all chemical shift differences, $|\nu_j - \nu_k|$, in a spin system are much greater than the corresponding coupling constants, $|J_{jk}|$, then the spectrum is first order, as noted in preceding sections. The spin system notation can therefore indicate if the spectrum is expected to be first order. This is clearly so for [**7**], and also for [**8**], which has an AMX spin system.

Conversely, if a chemical shift difference is of the order of, or less than, the corresponding coupling constant, adjacent letters of the alphabet are used for the two nuclei involved. Thus the two protons of [**9**] constitute an AB spin system. Note that this does not

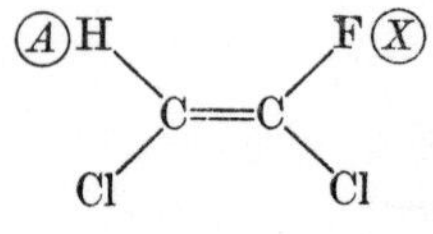

[**7**]

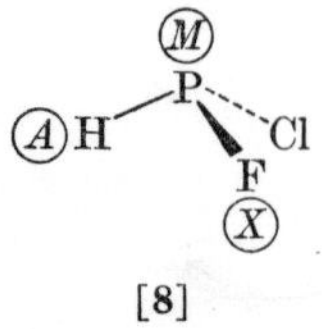

[**8**]

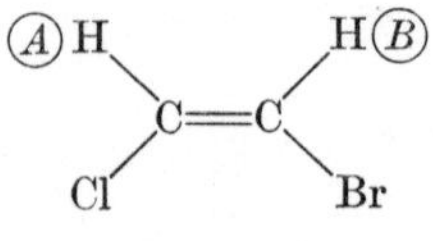

[**9**]

necessarily imply that the nuclei are close in space. If two nuclei, j and k, in a molecule are of the same isotope and it is not known if $|\nu_j - \nu_k| \gg J_{jk}$, it is usually safest to label the spin system as if $|\nu_j - \nu_k| \approx J_{jk}$, as has been done for [**9**] above. This implies that the spectrum may not be first order. However, there are several notations for indicating the type of symmetry of the spin system. It is usual to indicate magnetic equivalence by means of a numerical subscript; thus an A_2X system has two magnetically equivalent A nuclei (as for molecule [**1**]). Similarly the spin system of [**3**] is A_2X_2 and that of [**6**] is AB_4 or AX_4. In the notation of Haigh* square brackets are used to indicate repetition through symmetry of a group of nuclei. Thus molecule [**2**] has an $[AX]_2$ spin system. On occasion it is necessary to use more than one pair of square brackets for a given spin system. Thus molecule [**5**] possesses the system $[A[B]_2]_2$.

Ideally, any notation for spin systems should also specify the relationships between the coupling constants, since differences in these will affect the form of the spectrum. Thus there are two types of $[AX]_3$ system shown by the hypothetical molecular structures [**10**]

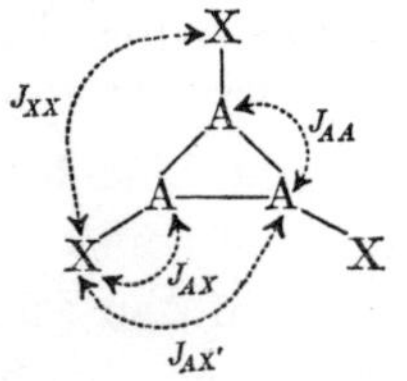

[**10**]

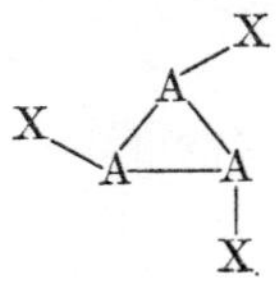

[**11**]

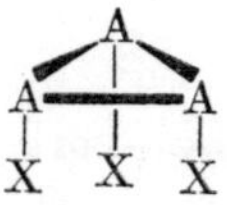

[**12**]

and [**11**], which have D_{3d} and C_{3h} symmetry respectively. There are only two types of (A, X) coupling constant for [**10**] whereas there are three for [**11**]. It should be emphasized that it is the relationships between spin parameters that matter (since they are scalar quantities), however, not necessarily the symmetry point group. Thus several related point groups may give the same spin system—for example, the C_{3v} atomic structure [**12**] has the same type of spectrum as that of the D_{3d} structure [**10**] because once more there are only two types of (A, X) coupling. When it is desired to discuss spin systems in detail it is necessary to distinguish on a molecular diagram (or a spin system diagram such as [**10**]) the various coupling constants and their notation. Slight difficulties are sometimes encountered if molecular diagrams are used, since it may not be possible to distinguish between

* Strictly speaking, Haigh's notation involves square brackets in cases [**1**], [**2**], and [**6**] which would be denoted $[A_2X]$, $[A_2X_2]$, and $[AB_4]$ spin systems respectively.

certain couplings by means of the spectrum alone. For instance there is an ambiguity between J_{ae} and J_{af} for [**5**]; the spectrum provides two different (A, B) coupling constants but does not indicate which is which. This arises because, as noted above, it is the spin system which matters for the spectrum, not the molecule itself. The notation $[A[B]_2]_2$ for [**5**] could indicate protons a, b, and e or protons a, b, and f as the repeated group $A[B]_2$. Normally such ambiguities may be resolved by consideration of values of coupling constants for related compounds.

1-15 First-order spectra involving magnetic equivalence

As stated in Section 1-13, spectra of spin systems with magnetic equivalence can be very simple. The circumstances under which this happens are analogous to those discussed earlier for first-order spectra, namely when all chemical shift differences between non-equivalent nuclei are much greater than the corresponding coupling constants between them. For this purpose, a set of magnetically equivalent nuclei is treated as a single group. The splitting patterns may be obtained as before, provided it is recognized that the equivalence of some nuclei implies that some coupling constants are equal, and therefore that some lines overlap. This is shown for the simplest case, that of the AX_2 spin system, in Fig. 1-17. This method of successive splittings can however be greatly simplified by recognizing that the overlapping of lines leads to the following rules:

(a) Splitting of a resonance due to coupling of the nucleus concerned with p magnetically equivalent nuclei leads to the appearance of $p + 1$ distinct lines.

(b) The intensity distribution within the $p + 1$ lines is that of the binomial coefficients, i.e. those of the expansion $(1 + x)^p$.

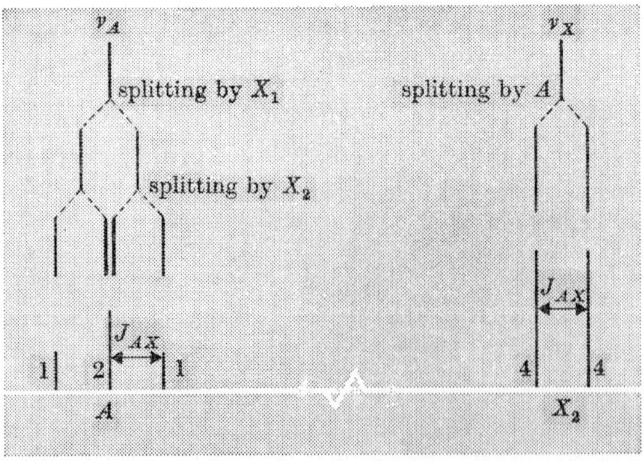

Fig. 1-17 The splitting pattern for an AX_2 spin system. The numbers adjacent to the lines give the intensities relative to that of the weakest A line.

The appropriate relative intensities for various values of p are given in Table 1-4. Three features need to be pointed out. The first is that if calculated intensities of lines in different parts of the spectrum are to be comparable, then account must be taken of the number of nuclei of each type. Thus in Fig. 1-17 the relative intensities are given in such a fashion that the total intensity in the X part is twice as great as the total intensity in the A part. Secondly, no account need be taken of coupling between nuclei that are magnetically equivalent. Such coupling will normally exist but it does not affect the spectrum, even in second-order cases. This can be rigorously

Table 1-4 Relative intensities within NMR multiplets due to coupling to p magnetically equivalent nuclei in first order

p	*Relative intensities (binomial coefficients)*
1	1:1
2	1:2:1
3	1:3:3:1
4	1:4:6:4:1
5	1:5:10:10:5:1
6	1:6:15:20:15:6:1

proved. Lastly, the splitting patterns need not be confused with those given by single nuclei of spin $>\frac{1}{2}$, since the intensity distributions are not the same even when the number of observed lines is identical. For instance, the A region of Fig. 1-17 is a 1:2:1 triplet because of coupling to two magnetically equivalent spin $\frac{1}{2}$ nuclei; coupling to a single spin 1 nucleus would have given a 1:1:1 triplet arising from the three equally probable m_I values for the spin 1 nucleus.

Figure 1-18 shows a spectrum of bromoethane at 100 MHz, which is of the first-order type. The CH_2 protons give rise to a 1:3:3:1 quartet because of coupling to the three methyl protons which are rendered magnetically equivalent by rapid internal rotation; likewise, the CH_3 protons give a 1:2:1 triplet (or 3:6:3 if considered relative to the 1:3:3:1 CH_2 quartet). The slight deviations from binomial intensity ratios are due to departure from first-order conditions.

1-16 Experimental aspects: the spectrometer

There are at the present time a variety of NMR spectrometers in use, mostly commercially made. There are instruments specially designed for broad-line work (i.e. measurements on solids) and for spin-echo

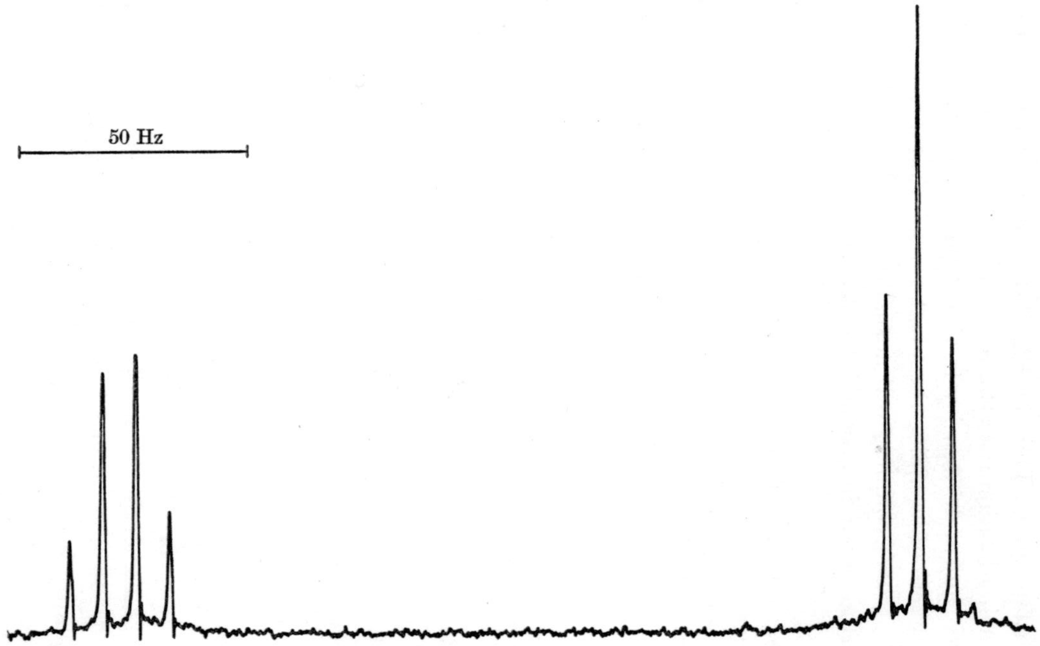

Fig. 1-18 100 MHz spectrum of bromoethane.

studies; these spectrometers will not be described. High-resolution NMR spectrometers, like any other spectroscopic instruments, require a source of radiation, a sample cell, a detector and a recording device, together with a method of scanning the spectrum. In addition, a magnet giving an extremely stable and homogeneous magnetic field is necessary. The range of magnetic fields normally used for proton resonance is 1·0 to 5·5 T, giving resonance frequencies from 40 to 220 MHz. The resolution normally required is better than 0·5 Hz, so at a proton frequency of 100 MHz the magnet homogeneity (which may be a limiting factor for the resolution) must be at least 1 part in 2×10^8, and the stability should be equally as good. The magnets themselves are products of precision engineering, but further devices are necessary to reach the required standard. One such device for obtaining good homogeneity is the use of shim (or Golay) coils. These are conducting loops situated at the pole faces which carry small variable currents, producing magnetic fields that can be adjusted to compensate for field gradients in the principal field. The magnets themselves may be of three types, viz. permanent magnets, ordinary electromagnets and superconducting electromagnets; the last-mentioned are the only type suitable for use above 2·5 T, but they are not yet common. The electromagnets have the advantage that the fields produced are far less sensitive to temperature changes than those from permanent magnets; however, the inherent field stability apart from the temperature effect is lower for electromagnets, they are costly to operate, and (except for the superconducting type) they require efficient water-cooling.

The source of electromagnetic radiation at the radio frequencies necessary is a crystal oscillator, which again must be of high stability. Usually its basic frequency is lower than that actually used in the spectrometer, and electronic circuits are used to provide appropriate frequency multiplication. In some spectrometers the frequency employed to obtain resonance is not that produced directly by multiplication from the crystal frequency (this is known as the centre-band frequency), but is a side band produced by modulation of the centre band using an audio frequency oscillator. Typical audio frequencies used are in the region of 2,000 Hz for proton resonance; the value may be varied linearly, thus providing a method of producing frequency-sweep spectra, while maintaining a constant magnetic field. Use of a side band in this fashion makes it possible to obtain added spectrometer stability by electronically compensating for changes in the centre band frequency (or that of a second side band) by changing the magnetic field (so as to keep the field-to-frequency ratio constant); such a technique is known as field-frequency locking, and is normally done using detection of an NMR signal at the second side band frequency to drive the field changes.

Throughout this book it will be assumed that spectra are obtained using the frequency-sweep mode. However, the resonance condition [Eqn. (1-18)] merely specifies a relationship between field and frequency. It is possible to obtain a spectrum by varying B_0 while keeping ν constant. This is known as the field-sweep mode, and is in fact the commonest method in use, since it is technically easy. The magnetic field strength is varied by employing auxiliary electric currents in conducting loops known as the sweep coils, situated at the pole faces. Because of the form of Eqn. (1-18), an increase of B_0 at constant ν to compensate for variations in σ is equivalent to a decrease of ν at constant B_0. Thus for a spectrum the term 'high frequency' (implying constant B_0) is the same as 'low field' (implying constant ν).

Physically, an NMR spectrometer usually consists of two parts: (a) the magnet with associated power supply and cooling unit (if it is an electromagnet), and (b) the operating console, which contains the radio frequency source and associated electronic circuits, most of the controls, and the recorder or recorders. However, the sample must be placed in the magnetic field, together with the actual irradiating and detecting devices. These features are located in a unit known as the probe, which is situated in the pole gap. The most commonly used system has separate coils for irradiation and detection and is hence known as the double-coil (or *induction*) method—the alternative

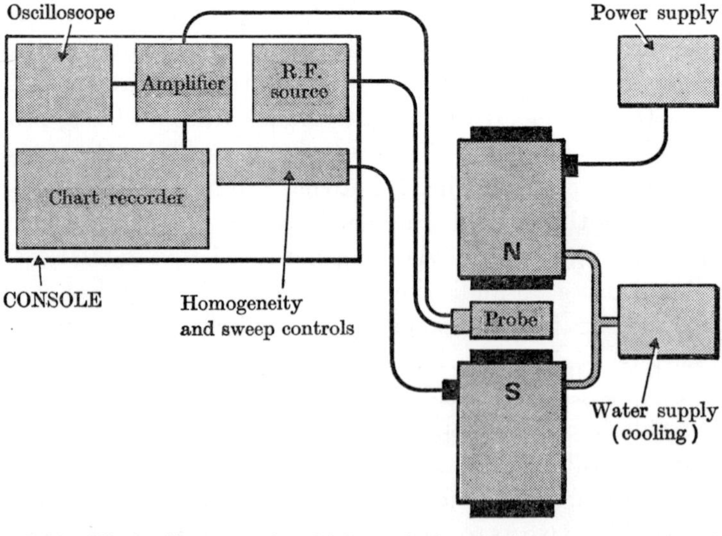

Fig. 1-19 Block diagram of a high-resolution NMR spectrometer using an electromagnet.

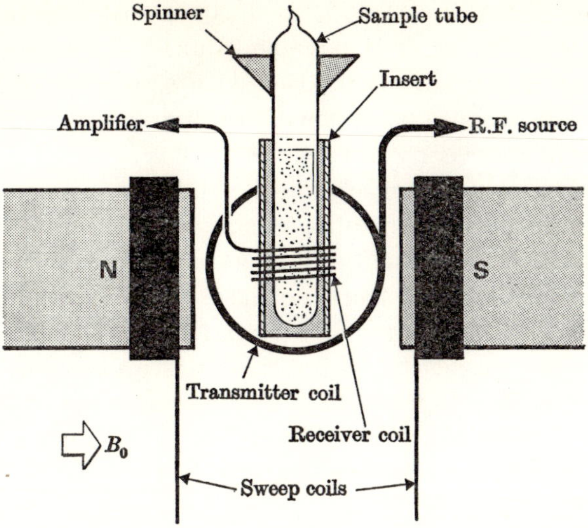

Fig. 1-20 The arrangements in the region of the pole gap of an NMR spectrometer (diagrammatic). The shim coils are not shown.

name is due to the mode of operation (a current is *induced* in the second coil when resonance occurs). As shown in Section 1-6, the irradiating frequency must correspond to a magnetic field $\mathbf{B}_1$ which is perpendicular to the primary field $\mathbf{B}_0$, but which oscillates with time (see also Chapter 4). The axis of the transmitter coil must therefore be perpendicular to $\mathbf{B}_0$; the axis of the receiver coil must be perpendicular to both $\mathbf{B}_0$ and $\mathbf{B}_1$, since there must be no direct induction of current in the receiver from the transmitter, only indirect induction via the magnetic resonance. The receiver coil is normally smaller in diameter than the transmitter coil and it is wound on a vertical tube of glass or non-magnetic metal, forming what is known as the insert. The small current induced in the receiver coil at resonance is amplified many times before being recorded. Figure 1-19 shows a block diagram of an NMR spectrometer using an electromagnet, while Fig. 1-20 illustrates diagrammatically the arrangements in the vicinity of the pole gap.

1-17 Experimental aspects: the sample and the spectrum

As explained previously, high-resolution NMR studies are made almost exclusively on fluids, usually liquids. Fortunately, glass does not give absorptions detectable under normal operation, so that sample cells are glass tubes (made of either Pyrex or soda glass) placed vertically in the insert (in the pole gap—see Fig. 1-20). The tubes may either be sealed at the top or left unsealed. The normal size of tube for proton resonance is 5 mm in outside diameter.

About 3 cm depth of sample is used, so that the volume required is ca. 0·5 cm^3; however, it is possible to use less sample (say ca. 0·05 cm^3) if a special microtube is used. Unfortunately, NMR is not a very sensitive technique, and with the normal tubes solution concentrations of at least 1 % are necessary. Except when solvent effects on coupling constants or chemical shifts are being studied, it is normal to use saturated solutions (care being taken to avoid having any solid particles) or neat* liquids. At least one atmosphere pressure is necessary for the NMR spectra of gases to be observed under normal conditions. The sensitivity of NMR (in terms of signal-to-noise ratio) increases as the magnetic field is increased; this is one of the principle reasons for the development of spectrometers using superconducting magnets. As explained in Section 1-9, a reference compound is normally added to the sample, either directly (internal standard) or in a separate capillary tube (external standard). Special sets of concentric tubes are available for the latter case, the outer tube being of the appropriate outside diameter.

The tubes are normally spun about the long axis at rates of ca. 30 Hz in order to average inhomogeneities of the magnetic field; this procedure results in an improvement of resolution. In order to do this the top of the tube is gripped in a plastic spinner; the spinning is affected by directing a current of air tangentially at the spinner. The spinning process often introduces unwanted spinning side bands into the spectrum. These are symmetrically placed on either side of a genuine resonance (see Fig. 1-9) and separated from it by the spinning rate (in Hz). In principle, there are further side bands at multiples of that separation from a genuine band. The spectrometer conditions are adjusted so as to minimize the intensity of spinning side bands.

For many experiments it is desirable to vary the temperature of the sample (while leaving that of the magnet fixed). This is accomplished by passing a stream of nitrogen gas into the insert past the sample tube; the gas is either pre-heated by passing it over an electrically heated wire coil, or pre-cooled, using a heat exchanger dipped in liquid nitrogen. The temperature is measured using a thermocouple, or, alternatively, making use of a calibrated temperature-dependent chemical shift difference.

The spectrum can be observed in two ways: either on an oscilloscope or on a paper chart. The oscilloscope is usually used only when adjusting the spectrometer conditions initially; the magnetic field is repetitively scanned so that its variation coincides with the travel

* The term 'neat' is used in order to avoid the connotations of 'pure,' particularly as a reference compound is normally added.

of the light spot across the oscilloscope screen. An absorption will therefore occur at the same position on the screen for every scan until conditions are altered. The spectrum may be permanently obtained using the chart recorder; conventionally the frequency increases from right to left on the charts. With modern stable spectrometers the charts may be pre-calibrated in Hz (or δ values); measurements are never made in field units even when field sweep is used. Alternatively, calibration may be carried out by generating side bands of a sharp resonance by means of an audio-frequency oscillator at a known frequency setting. By suitable adjustment of the spectrometer, the frequency scale of the chart may be expanded (so as to observe fine structure of bands clearly), and different regions of the spectrum may be brought within the range of the chart (by means of a field or frequency shift).

The intensity scale shows total absorption and not percentage absorption (as is the case for infrared spectroscopy), since there is never an appreciable proportion of the transmitted signal absorbed. Signal areas are therefore a direct measure of the amount of absorption, provided the spectrometer controls are not altered; this is the basis of the use of NMR to 'count' the relative numbers of different nuclei. When line widths are all the same, signal heights may be used as a measure of intensity. However, because of the effects of spin-spin coupling it is frequently not feasible to measure the total intensities attributed to given nuclei in a simple way. It may be noted that signal *areas* are really required; this is equivalent to an *integration* of the signal above the level of the base-line. It is normal practice to use automatic electronic integration, which may be done so as to trace on the spectrum a line whose height is proportional to the area of the signals scanned (see Fig. 1-9). Even with such a device, relative intensities cannot be obtained very accurately (only to ca. 5% usually); relative frequencies of absorption may be measured to a far greater accuracy ($\pm 0{\cdot}1$ Hz). Absolute values of these two parameters are not normally obtained. When dealing with fine-structure details, particularly when lines overlap, electronic integration is not of much use.

In order to obtain accurate values of relative absorption frequencies, it is necessary to have an expanded-scale spectrum and to scan the frequency (or field) very slowly—being careful, however, to avoid saturation (see Chapter 4). When a peak is scanned rapidly, three occurrences distort its shape (see Fig. 1-21). In the first place, the maximum of the peak is shifted to later in the scan, owing to the delay in response of the system (which does not have time to maintain equilibrium with the radiation as the scan advances). Secondly,

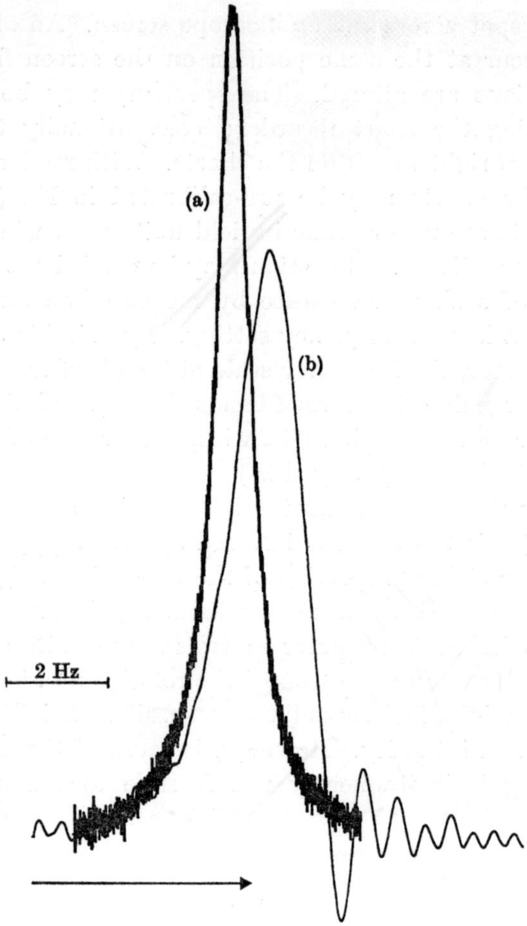

Fig. 1-21 The effect of scan rate on the shape of a resonance absorption: (a) 0·05 Hz s^{-1} (b) 2 Hz s^{-1}. The arrow indicates the direction of scan.

the maximum is lower, and thirdly the side of the peak that is scanned last exhibits the phenomenon of 'ringing' instead of a steady decay from maximum absorption (Fig. 1-21). To observe a lot of ringing during a rapid scan is a good sign, since this indicates that the magnetic field homogeneity is high. In fact, the ringing of a peak (on the oscilloscope) is often used as a monitor of homogeneity while adjusting the shim coils. Moreover, when a single peak is scanned the decay of the ringing should be exponential; departure from such a character in the decay is also an indication of a maladjustment of the shim coils. When a first-order multiplet is scanned rapidly, there is interference of the ringing of the different peaks and a beating pattern is obtained (Fig. 1-22). It can be shown that the time between beat maxima is $1/J$, and this fact may be used to measure the coupling constant, J, affecting the multiplet.

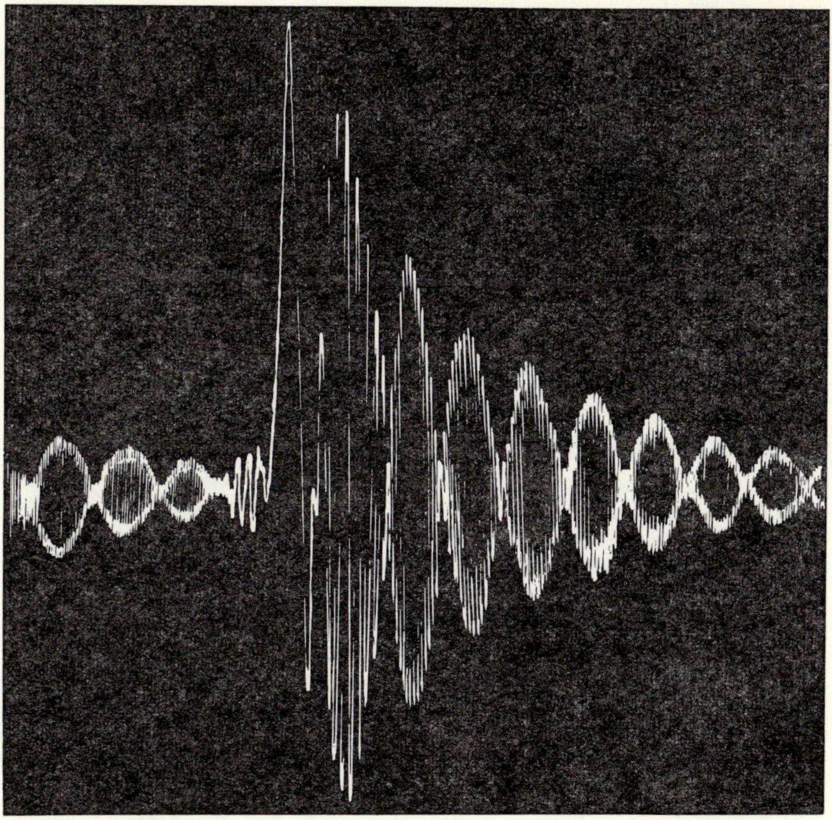

Fig. 1-22 The beating pattern obtained at 60 MHz when the methyl proton resonance of acetaldehyde is rapidly scanned (60 Hz s^{-1}) on a compressed scale. The separation of the beats is 0·35 s. The figure represents an oscilloscope trace which is repetitively scanned by the light spot. The small beats at the left are due to retention of 'memory' of the light spot, and are therefore spurious. (Based on photograph by courtesy of Perkin-Elmer Ltd.).

The fact that the peak maximum is shifted when the scan rate is finite implies an inaccuracy in measurement. Furthermore, temporary small variations in the magnetic field or frequency also produce spurious shifts. In the most accurate work it is usual to record and measure the spectrum several times (to compensate for the latter fault) and to sweep the frequency from high to low in some scans and from low to high in others (the error due to finite scan rate being reversed in direction by this procedure).

Problems

[N.B. Make use of Tables 1-1 and 1-2 in answering these questions.]

1-1 Calculate the magnetogyric ratio for ^{75}As, which has spin quantum number $I = 3/2$ and a maximum observable magnetic moment 1·4349 nuclear magnetons.

WADHAM COLLEGE
LIBRARY.

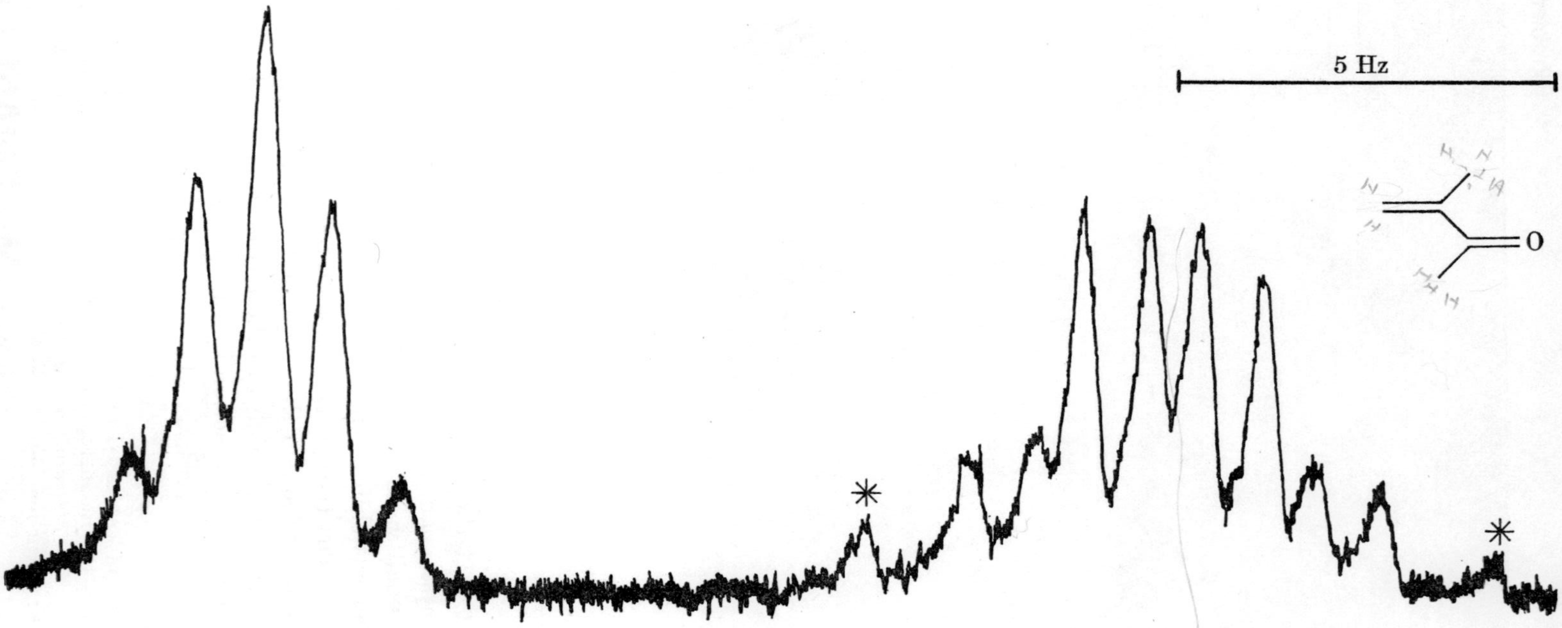

Fig. 1-23 The 60 MHz proton resonance spectrum for the ethylenic protons of methyl isopropenyl ketone, $CH_2{=}CMeCOMe$ (see Problem 1-5).

1-2 At what frequency do the following nuclei give NMR absorption for a magnetic field of 1·5 T: (a) ^{29}Si, (b) ^{199}Hg, (c) ^{35}Cl?

1-3 In a certain spectrometer, 'bare' protons absorb at 42·577 MHz. At what magnetic field does the spectrometer operate? To what value must the field be raised in order to achieve resonance for ^{31}P nuclei?

1-4 Derive Eqn. (1-21) from considerations similar to those used in Section 1-4.

1-5 Explain the appearance of the ethylenic part of the PMR spectrum of methyl isopropenyl ketone (Fig. 1-23, recorded at 60 MHz). Note that the small peaks marked with asterisks are due to impurities.

1-6 An AX spin system gives rise to absorption lines at 29·6, 34·1, 391·1, and 295·8 Hz to high frequency of the resonance of TMS for an instrument operating at 40 MHz. Give the δ values for the two types of proton. At what separations (in Hz) from TMS would resonance occur for an instrument operating at 60 MHz?

1-7 What is the difference between the proton resonance frequencies of benzene and acetone (7% in $CDCl_3$) at a constant magnetic field of $B_0 = 2{\cdot}35$ T? What is the difference in the shielding constant between the two molecules? Hence, what is the difference in the magnetic fields actually experienced by the protons of the two molecules.

1-8 (a) Re-label the transitions (α, β etc.) of Fig. 1-13 for the case of negative J_{13} (all the remaining coupling constants being positive).
(b) Re-label the transitions (A_1, A_2 etc.) of Fig. 1-12 for the case of negative J_{MX} so that the labels remain consistent with Fig. 1-14.

1-9 Furan-2-aldehyde [**13**] has the following spin parameters:

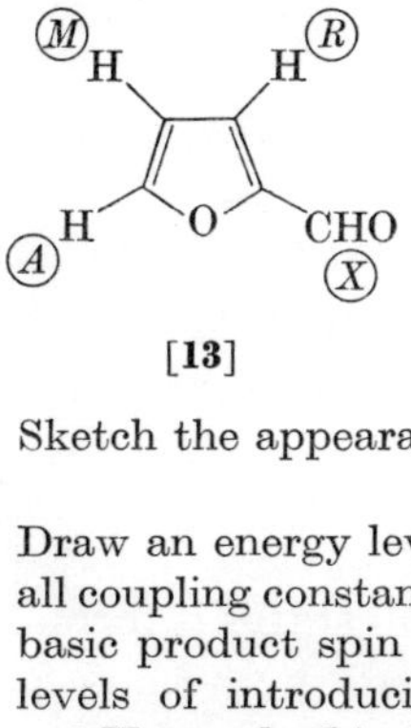

[**13**]

$$\begin{aligned}
\delta_A &= 7{\cdot}81 & J_{AM} &= 1{\cdot}65\ \text{Hz}\\
\delta_M &= 6{\cdot}67 & J_{MR} &= 3{\cdot}62\ \text{Hz}\\
\delta_R &= 7{\cdot}23 & J_{AR} &= 0{\cdot}80\ \text{Hz}\\
\delta_A &= 9{\cdot}66 & J_{AX} &= 0{\cdot}81\ \text{Hz}
\end{aligned}$$

$$J_{MX} \sim J_{RX} \sim 0$$

Sketch the appearance of each band of the spectrum in first order.

1-10 Draw an energy level diagram for a first-order three-spin system with all coupling constants zero. Clearly label the levels with the appropriate basic product spin wave functions. Indicate the effect on the energy levels of introducing (a) $J_{AM} = +10$ Hz, $J_{AX} = +6$ Hz, $J_{MX} = +2$ Hz and (b) $J_{AM} = -10$ Hz, $J_{AX} = -6$ Hz, $J_{MX} = -2$ Hz. Show that the *spectrum* is the same for (a) and (b), except for the labelling of the lines.

1-11 Suggest the notation to be used for the spin systems (for the nuclei with $I = \frac{1}{2}$ only) of the following molecules:
(a) *p*-bromochlorobenzene
(b) cyclopropyl chloride
(c) cyclohexane (as if fixed in a single chair form)
(d) SF_5CF_3
(e) *o*-chlorofluorobenzene
(f) PF_3Cl_2

1-12 Derive the first-order pattern of lines, with the intensities relative to the weakest line in the whole spectrum, for an $A_3M_4X_2$ spin system. Assume that $J_{AM} > J_{AX} > J_{MX} > 0$.

1-13 If each first-order transition has unit intensity, what is the total intensity in the A region of an A_pX_q spectrum?

2 Analysis of NMR spectra

2-1 Introduction

Values for coupling constants and chemical shifts can be obtained from first-order spectra by direct measurement of line positions, which are determined by the method of successive splittings described in Section 1-10. However, first-order spectra only occur if the relative chemical shift of any pair of nuclei is large compared with their mutual coupling. Such spectra may be recognized by the fact that the lines either have equal intensities, or show the binomial intensity patterns described in the previous chapter.

In this chapter we shall be concerned with the more complex spectra that are obtained when coupling constants and chemical shift differences are of comparable sizes. The practical problem of finding values of the spin parameters from an observed spectrum is best tackled by solving the converse problem—'given a set of J's and σ's what is the appearance of the spectrum?' As spectral lines arise from transitions between nuclear spin states, we must use quantum-mechanical methods to find the answer to this question.

The allowed energies and wave functions for molecules and atoms are found by solving the Schrödinger equation which may be written

$$\hat{\mathscr{H}}\psi = E\psi \tag{2-1}$$

The equation which determines the nuclear spin states is formally the same but $\hat{\mathscr{H}}$ and ψ are very different. Instead of involving the differential operators ∇^2 and functions of position, the Hamiltonian $\hat{\mathscr{H}}$ contains spin angular momentum operators. Consequently, the method of solution of the equation is quite different from the techniques used for the hydrogen atom. In fact we must go one step back towards the basic fundamentals of quantum mechanics and use the operator methods introduced by Dirac.

As quantum mechanics is concerned with predicting the results of experiments, observable quantities are particularly important. Every observable property corresponds to an operator, $\hat{A}$, and the results of the observation are deduced from the result of operating with $\hat{A}$ on the wave function. The latter contains all the information necessary for predicting the results of any observation. These often cannot be predicted exactly, and in such cases only the relative probabilities of different results can be forecast. The replacement of certainties by probabilities is one of the differences between classical and quantum systems.

Unfortunately, most chemists only meet quantum mechanics in the form of Schrödinger's equation. The operator techniques that must be used to understand spin systems are inherently simpler and involve less mathematics than the solution of the differential equation for the hydrogen atom. The only problem is unfamiliarity. The first sections of this chapter summarize the necessary background and definitions. The form of the spin Hamiltonian is given in Section 2-5, followed by a description of the calculation of the energy levels and spectra of one-, two- and many-spin systems (Sections 2-6 to 2-8). Finally, a few special techniques applicable to certain types of spectra are introduced.

2-2 Operators

The idea of an operator is frequently used in mathematics. An operator acts on a function to give another function.

$$\hat{A} f(x) = g(x) \tag{2-2}$$

We shall distinguish operators with a circumflex. There are many different types of operators; a few examples are:

(a) multiply by x: $\hat{x}$

$$\hat{x} f(x) = x f(x) \tag{2-3}$$

(b) differentiate with respect to x: $\dfrac{\hat{d}}{dx}$

$$\frac{\hat{d}}{dx} f(x) = f'(x) \tag{2-4}$$

(c) displace the origin of the coordinate system by a units: $\hat{T}$

$$\hat{T} f(x) = f(x - a) \tag{2-5}$$

Usually the new function g is different from f, but if it is only a multiple of f, so that

$$\hat{A} f(x) = a f(x) \tag{2-6}$$

then $f(x)$ is said to be an *eigenfunction* of $\hat{A}$ with *eigenvalue* a. For example $\exp kx$ is an eigenfunction of the operator $\dfrac{\hat{d}}{dx}$ with eigenvalue k.

$$\frac{\hat{d}}{dx} \exp kx = k \exp kx \tag{2-7}$$

As any value of k satisfies this equation, this operator has an infinite number of eigenvalues. Other operators, such as the spin operators discussed in Section 2-4, may only have a finite number of eigenfunctions. As the stationary states of a quantum mechanical system are eigenfunctions of the energy or Hamiltonian operator $\hat{\mathscr{H}}$, calculations of NMR spectra start with the problem of finding the eigenfunctions and eigenvalues of $\hat{\mathscr{H}}$.

The operators which correspond to observables in quantum mechanics are linear and Hermitian, that is, they satisfy the equations

$$\hat{A}[c_1 f_1(x) + c_2 f_2(x)] = c_1\hat{A} f_1(x) + c_2\hat{A} f_2(x) \tag{2-8}$$

and $$\int g^*(x)\hat{A} f(x)\, dx = \int f^*(x)\hat{A}\, g(x)\, dx \tag{2-9}$$

where $f^*(x)$ is the complex conjugate of $f(x)$ and $f(x)$, $g(x)$ are any functions. Two important consequences of Eqn. (2-9) which we shall use are, first, that the eigenvalues of Hermitian operators are real, and secondly that their eigenfunctions corresponding to different eigenvalues are orthogonal to each other.

The effect of a sum of operators operating on a function $f(x)$ is the sum of the functions produced by the individual operators. Thus, if

$$\hat{A}_1 f(x) = g_1(x) \qquad \text{and} \qquad \hat{A}_2 f(x) = g_2(x) \tag{2-10}$$

then $$(\hat{A}_1 + \hat{A}_2) f(x) = g_1(x) + g_2(x) \tag{2-11}$$

A product of operators operating on $f(x)$ is defined as successive operation by the operators starting with the one nearest the function.

$$\hat{A}_1\hat{A}_2 f(x) = \hat{A}_1\, g_2(x) = g_{12}(x) \tag{2-12}$$

The order in which the operators act is important, as the results of operating first with $\hat{A}_1$ and then with $\hat{A}_2$ is not necessarily the same as operating in the reverse order. For example

$$\frac{\hat{d}}{dx}\,\hat{x}\, f(x) \neq \hat{x}\,\frac{\hat{d}}{dx}\, f(x) \tag{2-13}$$

Two operators are said to commute if the order of operation makes no difference. The commutator $[\hat{A}_1, \hat{A}_2]$ of $\hat{A}_1$ and $\hat{A}_2$ is an operator defined to be

$$[\hat{A}_1, \hat{A}_2] = \hat{A}_1\hat{A}_2 - \hat{A}_2\hat{A}_1 \tag{2-14}$$

For example

$$\left[\hat{x}, \frac{\hat{d}}{dx}\right] = \hat{1} \tag{2-15}$$

where $\hat{1}$ is the unit operator which leaves any function unchanged.

2-3 Wave functions

The functions on which quantum-mechanical operators operate are known as wave functions and are functions of the variables of the particular system considered. For example, a particle in a box has a wave function which is a function of position (x, y, z). We shall use Dirac's notation for wave functions in this book because, although it may be less familiar, it is particularly suitable for spin functions. In this notation $|n\rangle$ is a wave function or ket and $\langle n|$ is its complex conjugate or bra. Whenever a bracket is completed integration over all variables is implied. Thus the normalization condition is

$$\langle n \mid n \rangle = 1 \tag{2-16}$$

and the expressions $\langle m| \hat{\mathscr{H}} |n\rangle$ and $\int \psi_m^* \hat{\mathscr{H}} \psi_n \, d\tau$ are equivalent.

The wave function itself cannot be observed, but the results of any observation can be predicted from it. If $|n\rangle$ is an eigenfunction of an operator $\hat{A}$

$$\hat{A} |n\rangle = a |n\rangle \tag{2-17}$$

then the result of the observation corresponding to $\hat{A}$ is always a. The wave function of the system is unaltered by this observation. If $|n\rangle$ is not an eigenfunction of $\hat{A}$ a further hypothesis must be introduced. Now $|n\rangle$ can always be written as a linear combination of eigenfunctions of $\hat{A}$. Suppose

$$|n\rangle = c_1 |a\rangle + c_2 |b\rangle \tag{2-18}$$

where $|a\rangle$ and $|b\rangle$ are normalized eigenfunctions of $\hat{A}$, with eigenvalues a and b. Then the hypothesis is that the probabilities of observing a and b are $c_1c_1^*$ and $c_2c_2^*$ respectively.† The mean or *expectation* value $\langle A \rangle$ for the observation corresponding to $\hat{A}$ is

$$\langle A \rangle = c_1c_1^*a + c_2c_2^*b \tag{2-19}$$

which can be written

$$\langle A \rangle = \langle n| \hat{A} |n\rangle \tag{2-20}$$

because eigenfunctions such as $|a\rangle$ and $|b\rangle$, which have different eigenvalues, are always orthogonal, that is $\langle a \mid b \rangle = \langle b \mid a \rangle = 0$. Equation (2-20) is universally true provided $|n\rangle$ is normalized.

A wave function may in certain circumstances be an eigenfunction of two (or more) different operators. If two operators commute, there are a complete set of functions which are simultaneously

† This hypothesis leads to the familiar interpretation of $\psi^*(x, y, z)\psi(x, y, z)\, dv$ as the probability of finding the particle in volume dv centred on (x, y, z).

eigenfunctions of both operators. On the other hand if they do not commute and

$$[\hat{A}, \hat{B}] = \hat{C} \tag{2-21}$$

then no functions are simultaneously eigenfunctions of $\hat{A}$ and $\hat{B}$ (unless they have eigenvalue zero for each operator and for the commutator $\hat{C}$). (See Problem 2-5.)

Expansion of a function in terms of a complete set

In Eqn. (2-16) the function $|n\rangle$ was expanded in terms of a number of eigenfunctions of an operator $\hat{A}$ in order to predict the possible results of the observation corresponding to $\hat{A}$. This technique of expansion in terms of a set of normalized and orthogonal functions is very useful, and we shall use it repeatedly in discussing spin functions. The set of functions chosen must satisfy certain conditions: they must be possible wave functions for the system (not necessarily stationary states) and there must be enough of them to make up the desired function. Often one uses a complete set which contains enough functions (sometimes an infinite number) to describe any possible wave function of the system as some linear combination of this basis set.

The coefficients in the expansion

$$|n\rangle = \sum_j c_j \, |j\rangle \tag{2-22}$$

can be found by completing the bracket with each $\langle j|$ in turn. As the set of functions $|j\rangle$ are orthogonalized and normalized

$$\langle j \mid n\rangle = c_j \tag{2-23}$$

the coefficient c_j is the overlap integral of $\langle j|$ and $|n\rangle$.

2-4 Spin operators and spin functions

The spin of an electron or a nucleus is an internal angular momentum of the particle. There are three independent components of spin angular momentum with corresponding operators $\hbar\hat{I}_x$, $\hbar\hat{I}_y$, and $\hbar\hat{I}_z$. These act on the spin wave functions of the particles which depend on the orientation of the particle and not on its position.

All the properties of the spin operators can be deduced from the commutation relations between the spin operators (where $i^2 = -1$). These are

$$\begin{aligned}[\hat{I}_x, \hat{I}_y] &= i\hat{I}_z \\ [\hat{I}_y, \hat{I}_z] &= i\hat{I}_x \\ [\hat{I}_z, \hat{I}_x] &= i\hat{I}_y\end{aligned} \tag{2-24}$$

An indication of how to derive these is given in text books on quantum mechanics.

The first deduction that we can make from these equations is that there are no states which are simultaneous eigenstates of $\hat{I}_x$, $\hat{I}_y$, and $\hat{I}_z$ (except states with zero spin angular momentum). This means that if a spin function is an eigenstate of $\hat{I}_z$ with eigenvalue m, the result of an observation of the z component of angular momentum can be predicted exactly, while the x and y components are uncertain (Problem 2-5). If two operators do commute it is possible to choose states which are simultaneously eigenstates of both operators. The result of each operation is its eigenvalue and the order of the operations is immaterial. Thus

$$\hat{A}\hat{B}\,|a, b\rangle = \hat{A}b\,|a, b\rangle = ab\,|a, b\rangle$$

and $$\hat{B}\hat{A}\,|a, b\rangle = \hat{B}a\,|a, b\rangle = ba\,|a, b\rangle \tag{2-25}$$

The square of the magnitude of the angular momentum corresponds to the operator $\hbar^2\hat{I}^2$ where

$$\hat{I}^2 = \hat{I}_x^2 + \hat{I}_y^2 + \hat{I}_z^2 \tag{2-26}$$

As this operator commutes with $\hat{I}_x$, $\hat{I}_y$, and $\hat{I}_z$ (Problem 2-6), it is possible to define both the magnitude of the spin angular momentum and its component in one direction simultaneously. Thus we may choose a set of normalized functions $|I, m\rangle$ which are eigenfunctions of both $\hat{I}^2$ and $\hat{I}_z$. These are the $2I + 1$ states with total angular momentum $\hbar[I(I + 1)]^{1/2}$ and with z component of angular momentum $\hbar m$ which were discussed in Chapter 1.

It is possible to find the allowed values of I and m from the commutation relations (2-25). This is done by introducing the raising and lowering operators $\hat{I}_+$ and $\hat{I}_-$ which convert the state $|I, m\rangle$ into $|I, m + 1\rangle$ and $|I, m - 1\rangle$ respectively. These operators are

$$\begin{aligned} \hat{I}_+ &= \hat{I}_x + i\hat{I}_y \\ \hat{I}_- &= \hat{I}_x - i\hat{I}_y \end{aligned} \tag{2-27}$$

which obey the commutation relations

$$\begin{aligned} [\hat{I}_+, \hat{I}_z] &= -\hat{I}_+ \\ [\hat{I}_-, \hat{I}_z] &= \hat{I}_- \\ [\hat{I}_+, \hat{I}_-] &= 2\hat{I}_z \end{aligned} \tag{2-28}$$

To show that $\hat{I}_+$ increases the m value by one unit, we use the first commutation relation acting on $|I, m\rangle$

$$(\hat{I}_+\hat{I}_z - \hat{I}_z\hat{I}_+)\,|I, m\rangle = -\hat{I}_+\,|I, m\rangle \tag{2-29}$$

but $\hat{I}_z \mid I, m\rangle = m \mid I, m\rangle$, so Eqn. (2-29) becomes

$$m\hat{I}_+ |I, m\rangle - \hat{I}_z\hat{I}_+ |I, m\rangle = -\hat{I}_+ |I, m\rangle \tag{2-30}$$

or $$\hat{I}_z(\hat{I}_+ |I, m\rangle) = (m + 1)(\hat{I}_+ |I, m\rangle) \tag{2-31}$$

This shows that the state $(\hat{I}_+ |I, m\rangle)$ is an eigenfunction of $\hat{I}_z$ with eigenvalue $(m + 1)$. It can also be shown that it is an eigenfunction of $\hat{I}^2$ with the same eigenvalue as $|I, m\rangle$, as $\hat{I}_+$ commutes with $\hat{I}^2$. However, as this new state is not necessarily normalized, we must write

$$\hat{I}_+ |I, m\rangle = c_{Im} |I, (m + 1)\rangle \tag{2-32}$$

where c_{Im} is a normalization factor which has the value $[I(I + 1) - m(m + 1)]^{1/2}$ (see Problem 2-8). Similarly

$$\hat{I}_- |I, m\rangle = [I(I + 1) - m(m - 1)]^{1/2} |I, m - 1\rangle \tag{2-33}$$

As the magnitude of a component of angular momentum cannot be greater than the total angular momentum, there must be a maximum value of m, $m_{\max}$, and $\hat{I}_+$ acting on this state must give zero.

$$\hat{I}_+ |I, m_{\max}\rangle = 0 \tag{2-34}$$

By considering the effect of operating with $\hat{I}^2$ on $|I, m_{\max}\rangle$ and $|I, m_{\min}\rangle$ it can be shown (Problem 2-7) that

$$\hat{I}^2 |I, m\rangle = m_{\max}(m_{\max} + 1) |I, m\rangle = m_{\min}(m_{\min} - 1) |I, m\rangle \tag{2-35}$$

As it must be possible to go from $|I, m_{\max}\rangle$ to $|I, m_{\min}\rangle$ by successive use of the lowering operator, $(m_{\max} - m_{\min})$ must be zero or a positive integer. The only possible solution is that

$$m_{\max} = -m_{\min} = I \tag{2-36}$$

where I has possible values $0, \frac{1}{2}, 1, \frac{3}{2}, \ldots$, and

$$\hat{I}^2 |I, m\rangle = I(I + 1) |I, m\rangle \tag{2-37}$$

All types of angular momentum are quantized in this way in quantum mechanics. Normally, different letters are used for different types of angular momentum, e.g. I for nuclear spin, S for electron spin. Orbital angular momenta have only integral quantum numbers (for example, s orbitals have zero angular momentum, p orbitals have $\ell = 1$, and d orbitals have $\ell = 2$).

Nuclei in their ground states have definite values of I which are tabulated in Table 1-1, and $2I + 1$ different energies in a magnetic field. The most important class of nuclei for nuclear resonance is that comprising those with quantum number $I = \frac{1}{2}$. Such nuclei have two states $|\frac{1}{2}, \frac{1}{2}\rangle$ and $|\frac{1}{2}, -\frac{1}{2}\rangle$ which are often designated $|\alpha\rangle$ and $|\beta\rangle$ (Section 1-11).

In the preceding discussion of angular momentum operators and their eigenvalues we did not have to specify the form of the operator or the coordinates of the wave function. All the information about eigenvalues and eigenfunctions was derived from the commutation relations. It is not necessary to know anything about the internal structure of nuclei in order to interpret nuclear resonance spectra. Nuclear physicists, however, relate the observed values of I and γ to the nuclear structure.

Summary of results

$$\hat{I}^2 |I, m\rangle = I(I + 1) |I, m\rangle \qquad I = 0, \tfrac{1}{2}, 1, \ldots$$
$$\hat{I}_z |I, m\rangle = m |I, m\rangle \qquad m = I, I - 1 \ldots -I$$
$$\hat{I}_+ |I, m\rangle = [I(I + 1) - m(m + 1)]^{1/2} |I, m + 1\rangle$$
$$\hat{I}_- |I, m\rangle = [I(I + 1) - m(m - 1)]^{1/2} |I, m - 1\rangle$$

For a particle with $I = \frac{1}{2}$ there are two states, designated α and β.

$$\alpha = |\tfrac{1}{2}, \tfrac{1}{2}\rangle \qquad \beta = |\tfrac{1}{2}, -\tfrac{1}{2}\rangle$$
$$\hat{I}^2 |\alpha\rangle = \tfrac{3}{4} |\alpha\rangle \qquad \hat{I}^2 |\beta\rangle = \tfrac{3}{4} |\beta\rangle$$
$$\hat{I}_z |\alpha\rangle = \tfrac{1}{2} |\alpha\rangle \qquad \hat{I}_z |\beta\rangle = -\tfrac{1}{2} |\beta\rangle$$
$$\hat{I}_+ |\alpha\rangle = 0 \qquad \hat{I}_+ |\beta\rangle = |\alpha\rangle$$
$$\hat{I}_- |\alpha\rangle = |\beta\rangle \qquad \hat{I}_- |\beta\rangle = 0$$

2-5 The nuclear spin Hamiltonian operator

The energy of a system is observable, and the corresponding operator $\hat{\mathscr{H}}$ is known as the Hamiltonian operator. It plays an important role in quantum mechanics, as its eigenfunctions describe the stationary states of the system. These are the solutions of the eigenvalue equation

$$\hat{\mathscr{H}}\psi = E\psi \tag{2-38}$$

The time-independent Schrödinger equation is a form of this equation.

The complete Hamiltonian for a molecule contains terms describing the kinetic energy of each particle, the potential energy of interactions between particles, and interactions of the particles with any external electric or magnetic fields. It can be divided approximately into terms representing electronic, vibrational, rotational, and nuclear spin contributions to the energy, and the wave function can be written as a product of electronic, vibrational, rotational, and nuclear spin functions, each of which are eigenfunctions of an effective Hamiltonian. The nuclear spin Hamiltonian contains nuclear spin operators and constants, the coupling constants and chemical shifts, whose values all depend on the electronic, vibrational and rotational states of the molecule. The nuclear spin states are eigenstates of this Hamiltonian.

The spin Hamiltonian used to interpret nuclear resonance spectra obtained from fluid samples is

$$\hat{\mathscr{H}} = -\sum_{j} (2\pi)^{-1} \gamma_j B_0(1 - \sigma_j)\hat{I}_{jz} + \sum_{j<k} J_{jk}\hat{\mathbf{I}}_j \cdot \hat{\mathbf{I}}_k \quad (2\text{-}39)$$

This equation has the dimensions of frequency and not energy. It is conventional and convenient in nuclear resonance to use $\hat{\mathscr{H}}$ for the energy operator divided by Planck's constant. The eigenfunctions are unchanged, but the differences in eigenvalues give transition frequencies directly. This Hamiltonian is the quantum-mechanical equivalent of the first-order equation of Section 1-12. The first term represents the interaction of the nuclear magnetic moments with the applied field $\mathbf{B}_0$. The energy of a nucleus in a magnetic field $\mathbf{B}$ is classically (Section 1-5)

$$U = -\boldsymbol{\mu} \cdot \mathbf{B} \quad (2\text{-}40)$$

As the magnetic dipole of a nucleus is related to its angular momentum by the magnetogyric ratio γ, the operator corresponding to the magnetic moment of nucleus j is

$$\hat{\boldsymbol{\mu}}_j = \gamma_j \hbar \hat{\mathbf{I}}_j \quad (2\text{-}41)$$

and the quantum-mechanical equivalent of Eqn. (2-40) is

$$h\hat{\mathscr{H}} = -\gamma_j \hbar \mathbf{B} \cdot \hat{\mathbf{I}}_j \quad (2\text{-}42)$$

The first term in Eqn. (2-39) is identical to this with coordinate axes chosen* with the z direction along $\mathbf{B}_0$ and B replaced by $B_0(1 - \sigma)$ to represent the shielding discussed in Section 1-8. The second term is the spin–spin coupling between different nuclei in the same molecule, discussed in Section 1-10.

The Hamiltonian (2-39) is appropriate for molecules which make frequent random collisions. Isolated molecules in a definite rotational and vibrational states (e.g. in a molecular beam), and molecules in solids, have more complex spin Hamiltonians. In a gas at a pressure of about 1 atmosphere or more, molecules frequently

* Some books use the opposite convention with the z axis in the $-\mathbf{B}_0$ direction. The first term in the Hamiltonian then has the opposite sign, but the spectrum is unchanged.

collide, changing their vibrational and rotational states and their relative positions in a random way; the observed spin Hamiltonian is the average over this motion. The values of J and σ are the mean values and show a small temperature variation due to changes of population of individual states. Normally, all terms in the spin Hamiltonian which depend on the relative orientations of the molecule and the applied field average to zero. In a liquid crystal solvent or in an electric field, however, some orientations are more probable than others and additional terms appear. The spin Hamiltonian for a solid also contains orientation-dependent terms, and, in addition, includes interactions between molecules.

2-6 The nuclear resonance spectrum of a single nucleus

The simplest type of nuclear resonance spectrum is that of a molecule containing a single magnetic nucleus A. The spin Hamiltonian is

$$\hat{\mathscr{H}} = -\nu_A \hat{I}_{Az} \tag{2-43}$$

where from the general spin Hamiltonian (2-39) the constant ν_A is

$$\nu_A = (2\pi)^{-1}\, \gamma_A B_0 (1 - \sigma_A) \tag{2-44}$$

The eigenstates of this Hamiltonian are the functions $|I, m\rangle$ with eigenvalues $-m\nu_A$. The energy level diagram consists of a series of $2I + 1$ equally spaced levels, as already described in Section 1-5 and shown in Fig. 1-2.

To predict the spectrum, the frequencies and relative probabilities of all possible transitions between states with different m values must be calculated. In the NMR spectrometer a radio frequency field $B_1 \cos \omega t$ is applied perpendicular to the static field B_0, say in the x direction. This adds a small time-dependent term to the Hamiltonian

$$\hat{\mathscr{H}}(t) = (2\pi)^{-1}\gamma B_1 \hat{I}_x \cos \omega t = \hat{\mathscr{H}}_1 \cos \omega t \tag{2-45}$$

which induces transitions between the states. The resulting probability per unit time of a spin in state $|m\rangle$ changing to state $|n\rangle$ can be found from time-dependent perturbation theory. It is

$$(2\pi)^2 \,|\, \langle n|\, \hat{\mathscr{H}}_1 \,|m\rangle \,|^2 \delta(E_n - E_m - h\nu) \tag{2-46}$$

where $\delta(x)$ is one if x is zero and zero otherwise. The rate of absorption of energy is Eqn. (2-46) multiplied by the size of the quantum, ΔE_{nm}, and the difference in the number of spins in the two states. The final expression is proportional to

$$\frac{\gamma^2 B_1^2 (\Delta E_{nm})^2}{kT}\, |\langle n|\, \hat{I}_x \,|m\rangle|^2 \tag{2-47}$$

as the energy difference ΔE_{nm} is small compared to kT and $\hat{\mathscr{H}}_1$ is given by Eqn. (2-45).

The selection rules and relative transition probabilities can be most readily derived by rewriting $\hat{I}_x$ in terms of the raising and lowering operators

$$\langle n|\,\hat{I}_x\,|m\rangle = \tfrac{1}{2}\langle n|\,\hat{I}_+ + \hat{I}_-\,|m\rangle \tag{2-48}$$

As $\hat{I}_+$ and $\hat{I}_-$ raise and lower the m value by one unit, the selection rule is

$$\Delta m = \pm 1 \tag{2-49}$$

The energy levels are in the order of their m values with maximum m at minimum energy. Net absorption of energy occurs when m decreases, so the selection rule is $\Delta m = -1$. Such transitions are degenerate for a single nucleus and the spectrum consists of a single line at ν_A.

2-7 Spectra of molecules with two coupled nuclei

For the remainder of the this chapter we shall confine our attention to nuclei of spin $\frac{1}{2}$. Using the notation discussed in Section 1-14 a molecule with two spin $\frac{1}{2}$ nuclei contains an AB nuclear spin system. The Hamiltonian for this system is

$$\hat{\mathscr{H}}_{AB} = -\nu_A\hat{I}_{Az} - \nu_B\hat{I}_{Bz} + J\hat{\mathbf{I}}_A\,.\,\hat{\mathbf{I}}_B \tag{2-50}$$

where ν_A and ν_B are defined by Eqn. (2-44). The scalar product of the spin operators can be expanded to give

$$\hat{\mathscr{H}}_{AB} = -\nu_A\hat{I}_{Az} - \nu_B\hat{I}_{Bz} + J\hat{I}_{Az}\hat{I}_{Bz} + \tfrac{1}{2}J(\hat{I}_{A+}\hat{I}_{B-} + \hat{I}_{A-}\hat{I}_{B+}) \tag{2-51}$$

The spin states of this system can be written as linear combinations of the simple product states $|\alpha\alpha\rangle$, $|\alpha\beta\rangle$, $|\beta\alpha\rangle$, and $|\beta\beta\rangle$ introduced in Section 1-11. The effect of an A spin operator on one of these states depends on the first symbol, while the B spin operator acts on the second symbol, leaving the first unchanged.

For example
$$\hat{I}_{Az}\,|\,\alpha\beta\rangle = \tfrac{1}{2}\,|\alpha\beta\rangle \tag{2-52}$$

$$\hat{I}_{B+}\,|\alpha\beta\rangle = |\alpha\alpha\rangle \tag{2-53}$$

The problem of finding the linear combinations of these basis states which are stationary states is simplified by finding operators which commute with the Hamiltonian, as the stationary states are also eigenstates of any operators which do so (Section 2-3). One such operator is $\hat{F}_z$, the z component of the total spin angular momentum.

$$\hat{F}_z = \hat{I}_{Az} + \hat{I}_{Bz} \tag{2-54}$$

Both $\hat{I}_{Az}$ and $\hat{I}_{Bz}$ commute with the first three terms of Eqn. (2-51) as these contain $\hat{I}_z$ operators only. The commutators of $\hat{I}_{Az}$ and $\hat{I}_{Bz}$ with the fourth term are

$$[\hat{I}_{Az}, \hat{I}_{A+}\hat{I}_{B-}] = \hat{I}_{A+}\hat{I}_{B-} \quad \text{and} \quad [\hat{I}_{Bz}, \hat{I}_{A+}\hat{I}_{B-}] = -\hat{I}_{A+}\hat{I}_{B-} \tag{2-55}$$

As these are equal and opposite the sum, $\hat{F}_z$, commutes with the complete Hamiltonian, and the stationary states are also eigenstates of $\hat{F}_z$ with eigenvalues m_T. Table 2-1 shows the m_T values of the product functions which we are using as basis functions.

Table 2-1 m_T **values of product functions for two nuclei**

Function	m_T
$\alpha\alpha$	1
$\alpha\beta$	0
$\beta\alpha$	0
$\beta\beta$	-1

As $\alpha\alpha$ and $\beta\beta$ are the only functions with m_T values of 1 and -1 respectively, they must be stationary states. Their energies are found from the first three terms of the Hamiltonian

$$\hat{\mathscr{H}}\,|\alpha\alpha\rangle = [-\tfrac{1}{2}(\nu_A + \nu_B) + \tfrac{1}{4}J]\,|\alpha\alpha\rangle \tag{2-56}$$

$$\hat{\mathscr{H}}\,|\beta\beta\rangle = [+\tfrac{1}{2}(\nu_A + \nu_B) + \tfrac{1}{4}J]\,|\beta\beta\rangle \tag{2-57}$$

The functions $\alpha\beta$ and $\beta\alpha$ have the same value of m_T. They are not stationary states, as they are mixed by the last term of the Hamiltonian.

$$\hat{\mathscr{H}}\,|\alpha\beta\rangle = [-\tfrac{1}{2}(\nu_A - \nu_B) - \tfrac{1}{4}J]\,|\alpha\beta\rangle + \tfrac{1}{2}J\,|\beta\alpha\rangle \tag{2-58}$$

$$\hat{\mathscr{H}}\,|\beta\alpha\rangle = [\tfrac{1}{2}(\nu_A - \nu_B) - \tfrac{1}{4}J]\,|\beta\alpha\rangle + \tfrac{1}{2}J\,|\alpha\beta\rangle \tag{2-59}$$

The stationary states are combinations $c_1\,|\alpha\beta\rangle + c_2\,|\beta\alpha\rangle$ which satisfy the equation

$$\hat{\mathscr{H}}(c_1\,|\alpha\beta\rangle + c_2\,|\beta\alpha\rangle) = E(c_1\,|\alpha\beta\rangle + c_2\,|\beta\alpha\rangle) \tag{2-60}$$

If we write

$$\hat{\mathscr{H}}\,|\alpha\beta\rangle = \mathscr{H}_{11}\,|\alpha\beta\rangle + \mathscr{H}_{21}\,|\beta\alpha\rangle, \tag{2-61}$$

$\mathscr{H}_{11}$ and $\mathscr{H}_{21}$ are the integrals

$$\mathscr{H}_{11} = \langle\alpha\beta|\,\hat{\mathscr{H}}\,|\alpha\beta\rangle,\ \mathscr{H}_{21} = \langle\beta\alpha|\,\hat{\mathscr{H}}\,|\alpha\beta\rangle \tag{2-62}$$

which are often known as diagonal and off-diagonal matrix elements of $\hat{\mathscr{H}}$. Equation (2-60) becomes

$$[\mathscr{H}_{11}c_1 + \mathscr{H}_{12}c_2]\,|\alpha\beta\rangle + [\mathscr{H}_{21}c_1 + \mathscr{H}_{22}c_2]\,|\beta\alpha\rangle - E(c_1\,|\alpha\beta\rangle + c_2\,|\beta\alpha\rangle) = 0 \quad (2\text{-}63)$$

This equation is satisfied if the coefficients of the functions $|\alpha\beta\rangle$ and $|\beta\alpha\rangle$ are separately zero. This gives simultaneous equations for c_1 and c_2 (the secular equations)

$$\begin{aligned}(\mathscr{H}_{11} - E)c_1 + \mathscr{H}_{12}c_2 &= 0\\ \mathscr{H}_{21}c_1 + (\mathscr{H}_{22} - E)c_2 &= 0\end{aligned} \quad (2\text{-}64)$$

which can be written in matrix form

$$\begin{bmatrix}\mathscr{H}_{11} - E & \mathscr{H}_{12}\\ \mathscr{H}_{21} & \mathscr{H}_{22} - E\end{bmatrix}\begin{bmatrix}c_1\\ c_2\end{bmatrix} = 0 \quad (2\text{-}65)$$

These equations are only consistent (two equations, one unknown c_1/c_2) if the determinant

$$\begin{vmatrix}\mathscr{H}_{11} - E & \mathscr{H}_{12}\\ \mathscr{H}_{21} & \mathscr{H}_{22} - E\end{vmatrix} = 0 \quad (2\text{-}66)$$

i.e. $(\mathscr{H}_{11} - E)(\mathscr{H}_{22} - E) - \mathscr{H}_{12}\mathscr{H}_{21} = 0$ (2-67)

The two values of E for which this secular determinant is zero are the two allowed energy levels, with energies (using $\mathscr{H}_{12} = \mathscr{H}_{21}$) given by

$$E_1 = \tfrac{1}{2}(\mathscr{H}_{11} + \mathscr{H}_{22}) + \tfrac{1}{2}[(\mathscr{H}_{11} - \mathscr{H}_{22})^2 + 4\mathscr{H}_{12}^2]^{1/2} \quad (2\text{-}68)$$

$$E_2 = \tfrac{1}{2}(\mathscr{H}_{11} + \mathscr{H}_{22}) - \tfrac{1}{2}[(\mathscr{H}_{11} - \mathscr{H}_{22})^2 + 4\mathscr{H}_{12}^2]^{1/2} \quad (2\text{-}69)$$

The spin functions corresponding to these two values of E are now found by substituting in the secular equations (2-65), and using the subsidiary condition that the function is normalized

$$c_1^2 + c_2^2 = 1 \quad (2\text{-}70)$$

The two spin functions are distinguished by an additional subscript. The eigenvector $[c_{11}\ c_{12}]$ satisfies Eqn. (2-60) when $E = E_1$, while $[c_{21}\ c_{22}]$ is the solution when $E = E_2$. The normalization condition may be incorporated by defining an angle of θ such that $c_{11} = \cos\theta$, $c_{12} = \sin\theta$. As $[c_{21}\ c_{22}]$ must be orthogonal to $[c_{11}\ c_{12}]$, $c_{21} = -c_{12}$; $c_{11} = c_{22}$. The value of θ for the AB system is given by

$$\begin{aligned}\sin 2\theta &= J/[\nu_A - \nu_B)^2 + J^2]^{1/2}\\ \cos 2\theta &= (\nu_A - \nu_B)/[(\nu_A - \nu_B)^2 + J^2]^{1/2}\end{aligned} \quad (2\text{-}71)$$

Table 2-2 Stationary states of the *AB* system

Spin function[(a)]	*Energy*
$\alpha\alpha$	$-\frac{1}{2}(\nu_A + \nu_B) + \frac{1}{4}J$
$\cos\theta\ \alpha\beta + \sin\theta\ \beta\alpha$	$-\frac{1}{4}J + \frac{1}{2}[(\nu_A - \nu_B)^2 + J^2]^{1/2}$
$-\sin\theta\ \alpha\beta + \cos\theta\ \beta\alpha$	$-\frac{1}{4}J - \frac{1}{2}[(\nu_A - \nu_B)^2 + J^2]^{1/2}$
$\beta\beta$	$\frac{1}{2}(\nu_A + \nu_B) + \frac{1}{4}J$

(a) $\sin 2\theta = J/[(\nu_A - \nu_B)^2 + J^2]^{1/2}$.

Table 2-2 shows the spin functions and energies of the stationary states of the AB system.

One extreme of the AB system is the AX system where $|\nu_A - \nu_B|$ is very large compared to $|J|$. It can be seen from Eqn. (2-71) that this corresponds to $\theta = 0$. In this case the functions $\alpha\beta$ and $\beta\alpha$ are not mixed, as the last term of the Hamiltonian is negligible compared to the difference $\mathscr{H}_{11} - \mathscr{H}_{22}$ of the diagonal elements. The term $\frac{1}{2}J(\hat{I}_{A+}\hat{I}_{B-} + \hat{I}_{B+}\hat{I}_{A-})$ can therefore be omitted from the calculation of an AX spectrum and of all first-order spectra (see Section 1-12).

As J increases, θ increases and the functions are mixed. The energy levels are separated by more than if there were no mixing. The other extreme of the AB system is reached when $\nu_A = \nu_B$, the A_2 system. Now θ is 45° and there is maximum mixing to give one state symmetric with respect to exchange of nuclei, $2^{-1/2}(\alpha\beta + \beta\alpha)$, with energy $\frac{1}{4}J$ and an anti-symmetric state $2^{-1/2}(\alpha\beta - \beta\alpha)$ with energy $-\frac{3}{4}J$. (Problem 2-11).

The spectrum

The intensity of a transition from state $|m\rangle$ to state $|n\rangle$ depends on the value of [cf. Eqn. (2-47)]

$$|\langle n|\ \gamma_A\hat{I}_{A-} + \gamma_B\hat{I}_{B-}\ |m\rangle|^2\ \Delta E^2_{nm} \tag{2-72}$$

If A and B are the same type of nucleus with the same magnetogyric ratio, the relative intensities of the transitions in the spectrum are given by

$$|\langle n|\ \hat{F}_-\ |m\rangle|^2 \tag{2-73}$$

where $\hat{F}_-(= \hat{I}_{A-} + \hat{I}_{B-})$ lowers the value of m_T by one unit, giving a selection rule

$$\Delta m_T = -1 \tag{2-74}$$

The AB system has four allowed transitions which are shown in the energy level diagram (Fig. 2-1).

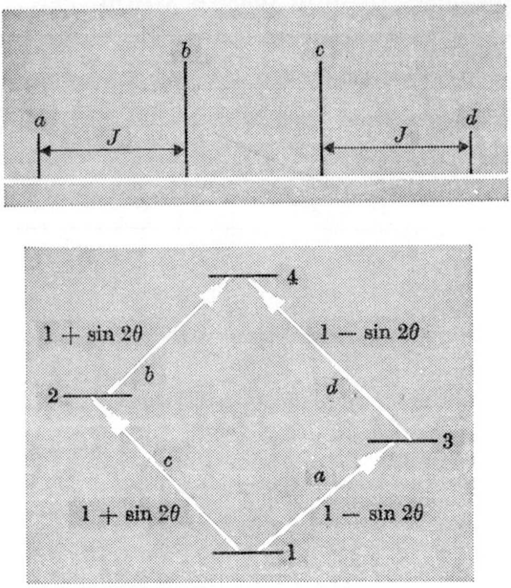

Fig. 2-1 Energy level diagram and spectrum of an *AB* spin system. Corresponding transitions in the spectrum and energy level diagram have been labelled, and intensities are shown on the energy level diagram.

Their relative intensities are found by operating with $\hat{F}_-$ on the lower state. For example

$$\hat{F}_- |2\rangle = \cos\theta \hat{F}_- |\alpha\beta\rangle + \sin\theta \hat{F}_- |\beta\alpha\rangle \tag{2-75}$$

but $$\hat{F}_- |\alpha\beta\rangle = \hat{I}_{A-} |\alpha\beta\rangle + \hat{I}_{B-} |\beta\alpha\rangle$$

$$= |\beta\beta\rangle + 0 \tag{2-76}$$

so $$\langle 4| \hat{F}_- |2\rangle = (\cos\theta + \sin\theta) \tag{2-77}$$

and the intensity of this transition is proportional to $(1 + \sin 2\theta)$.

Table 2-3 shows the transition frequencies and intensities of the *AB* system. Figure 2-2 shows the way in which the *AB* spectrum varies as θ increases. When $J/(\nu_A - \nu_B)$ is zero (the *AX* limit), the spectrum consists of four lines of equal intensity. Lines *c* and *d* arise from transitions in which the spin of the *A* nucleus changes and that of the *X* nucleus is unaltered. These can be described accurately

Table 2-3 NMR spectrum of the *AB* system

Transition		*Frequency*[(a)]	*Relative Intensity*
d	$4 \leftarrow 3$	$\frac{1}{2}(\nu_A + \nu_B) + \frac{1}{2}J + \frac{1}{2}D$	$1 - \sin 2\theta$
b	$4 \leftarrow 2$	$\frac{1}{2}(\nu_A + \nu_B) + \frac{1}{2}J - \frac{1}{2}D$	$1 + \sin 2\theta$
c	$2 \leftarrow 1$	$\frac{1}{2}(\nu_A + \nu_B) - \frac{1}{2}J + \frac{1}{2}D$	$1 + \sin 2\theta$
a	$3 \leftarrow 1$	$\frac{1}{2}(\nu_A + \nu_B) - \frac{1}{2}J - \frac{1}{2}D$	$1 - \sin 2\theta$

(a) $D^2 = [(\nu_A - \nu_B)^2 + J^2]$; $\sin 2\theta = J/D$.

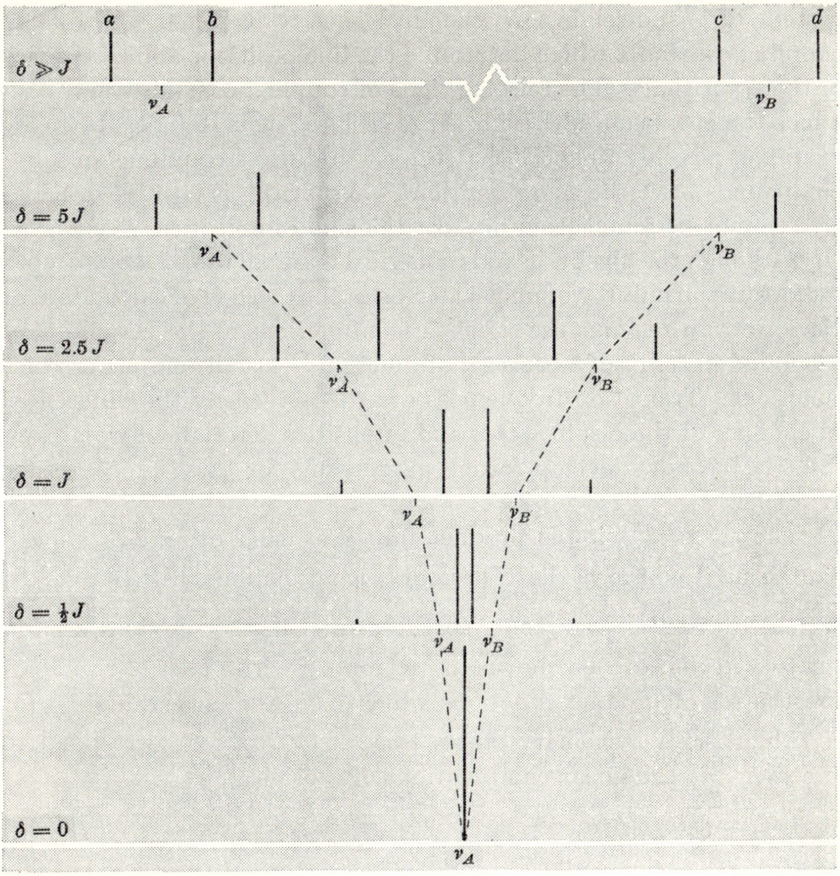

Fig. 2-2 AB spectra showing the effect of varying the ratio of J, the coupling constant, to δ, the difference in A and B nuclear resonance frequencies. The first spectrum is an AX type and the last is an A_2 type.

as A lines, and are centred on ν_A and split by J. This is the first-order situation described in Chapter 1.

As $|J/(\nu_A - \nu_B)|$ increases, the states are mixed and the line intensities are no longer equal. Furthermore, lines c and d are no longer strictly A transitions. In this system, J is given exactly by the separation of lines c and d (or a and b), but this is not generally true in systems with more nuclei. The value of $|\nu_A - \nu_B|$ is the square root of the product of line separations $(ad)(bc)$. In the A_2 limit the inner lines coalesce at ν_A, while the outer lines (a and d) have zero intensity. Thus in this limit J cannot be measured.

2-8 Systems with more than two nuclei

The spectra of systems with more than two nuclei become increasingly complex, with more transitions involving more parameters—a system of N nuclei has N chemical shift values and $\frac{1}{2}N(N-1)$ coupling constants which determine the line positions and intensities of the transitions. The relative signs of coupling constants may also affect the spectrum, although the 'absolute' signs do not. It is, for example, possible to show that the *cis* and *trans* couplings in vinyl compounds have the same sign, but whether this is positive or negative cannot be found from line positions and intensities.

Although the algebraic and numerical manipulation becomes more tedious as the number of nuclei increases, the method of calculation of the spectrum remains the same. Eigenfunctions of the Hamiltonian are found in terms of a basis set of functions—usually simple product functions. Transition frequencies are then given by the differences in energies of these spin states and intensities determined from Eqn. (2-46). We now develop formal expressions for these processes for N nuclei.

The set of 2^N simple product functions such as $|\alpha_1\alpha_2\beta_3 \ldots \alpha_N\rangle$ can be used as the basis. These are eigenfunctions of $\hat{F}_z(\hat{\mathbf{F}} = \sum_j \hat{\mathbf{I}}_j)$ which, by similar arguments to those used in the previous section for two nuclei, can be shown to commute with $\hat{\mathscr{H}}$. If $|\phi_1\rangle \ldots |\phi_n\rangle$ are the set of product functions which have the eigenvalue m_T for $\hat{F}_z$, then the n stationary states $|\psi_1\rangle \ldots |\psi_n\rangle$ which correspond to the same m_T are linear combinations of these $|\phi_j\rangle$.

$$|\psi_k\rangle = \sum_j c_{kj}\,|\phi_j\rangle \tag{2-78}$$

The stationary states $|\psi_k\rangle$ are eigenstates of $\hat{\mathscr{H}}$

$$\hat{\mathscr{H}}\,|\psi_k\rangle = E_k\,|\psi_k\rangle \tag{2-79}$$

which becomes

$$\sum_j \hat{\mathscr{H}} c_{kj}\,|\phi_j\rangle = \sum_j E_k c_{kj} = |\phi_j\rangle \tag{2-80}$$

when written in terms of the basis functions. Completing the brackets with each $\langle\phi_\ell|$ in turn we obtain equations such as

$$\sum_j \langle\phi_\ell\,|\mathscr{H}|\,\phi_j\rangle\, c_{kj} = E_k c_{k\ell} \tag{2-81}$$

These are the secular equations which can be written in matrix form

$$\begin{bmatrix} \mathscr{H}_{11} - E & \mathscr{H}_{12} & \cdots \\ \mathscr{H}_{21} & \mathscr{H}_{22} - E & \\ \cdot & & \\ \cdot & & \\ \cdot & & \end{bmatrix} \begin{bmatrix} c_1 \\ c_2 \\ \cdot \\ \cdot \\ \cdot \end{bmatrix} = 0 \tag{2-82}$$

where we have dropped the subscript k. They are consistent only if the secular determinant

$$\begin{vmatrix} \mathscr{H}_{11} - E & \mathscr{H}_{12} & \cdots \\ \mathscr{H}_{21} & \mathscr{H}_{22} - E & \\ \cdot & & \\ \cdot & & \\ \cdot & & \end{vmatrix} = 0; \tag{2-83}$$

solving this nth order equation for E gives n possible values of E and n spin eigenfunctions. The process is repeated for every value of m_T.

This process of solving a set of secular equations is often referred to as 'diagonalizing the Hamiltonian'. If a complete set of functions ($|\phi_1\rangle \ldots |\phi_n\rangle$) for a problem is given, then all the information about the Hamiltonian is contained in the matrix

$$\begin{bmatrix} \mathscr{H}_{11} & \mathscr{H}_{12} & \cdots \\ \mathscr{H}_{21} & \mathscr{H}_{22} & \\ \cdot & & \\ \cdot & & \\ \cdot & & \end{bmatrix} \tag{2-84}$$

which is said to represent the Hamiltonian in this basis. If another basis is chosen, then the form of the matrix changes; if eigenstates ($|\psi_1\rangle \ldots |\psi_n\rangle$) are chosen as basis there are no off-diagonal elements

$$\begin{bmatrix} \mathscr{H}_{11} & 0 & 0 & \cdots \\ 0 & \mathscr{H}_{22} & 0 & \\ 0 & 0 & \mathscr{H}_{33} & \cdots \\ \cdot & & & \\ \cdot & & & \\ \cdot & & & \end{bmatrix} \tag{2-85}$$

and the Hamiltonian is diagonal. Diagonalizing the Hamiltonian is not always easy. If the basis is the set of simple product functions, the Hamiltonian matrix consists of a number of blocks or sub-matrices, each of which corresponds to a different value of m_T (Fig. 2-3). If any of these blocks is bigger than 2×2, it cannot normally be diagonalized explicitly in terms of J's and ν's. Instead, numerical values of the constants are inserted and the eigenvalues found by an iterative process. This is tedious, but can conveniently be performed by a computer.

In systems where some nuclei are equivalent it is possible to reduce the size of the sub-matrices further. If a group of nuclei are

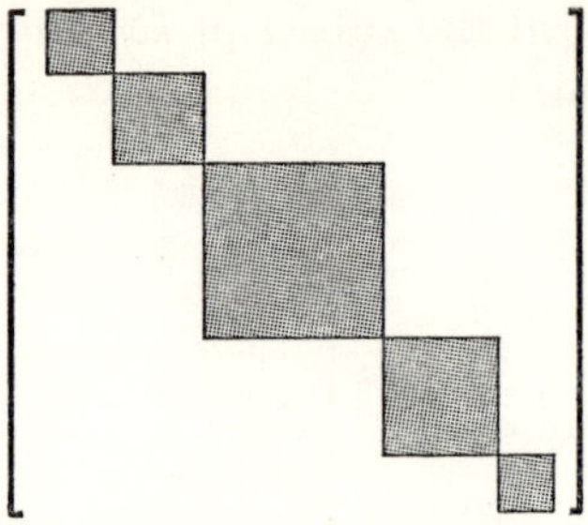

Fig. 2-3 Form of Hamiltonian matrix when factorized into sub-matrices corresponding to different values of m_T.

magnetically equivalent, they can be replaced by a single 'composite particle', as described in Section 2-10. As chemically equivalent nuclei are related by symmetry operations of the point group of the molecule, group theoretical methods can be used to construct a basis of symmetry functions in a similar way to that used in constructing vibrational normal modes of molecules or molecular orbitals. Further details are given in the books by Pople, Schneider and Bernstein and by Emsley, Feeney, and Sutcliffe (Vol. 1). For details, see 'Suggestions for Further Reading'.

Once the Hamiltonian has been diagonalized and the energies and spin functions of the stationary states are known, the intensities of the transitions must be calculated. If the nuclei all have the same magnetogyric ratios, the relative intensities are given by the values of

$$|\langle n| \hat{F}_- |m\rangle|^2 \tag{2-86}$$

Nuclei with different magnetogyric ratios are discussed in the next section.

2-9 The X approximation

The resonance frequencies of nuclei with different magnetogyric ratios, such as ^{13}C and ^{31}P, always differ by a large amount compared to the coupling between them. We shall show in this section that transitions in such systems can correctly be described as ^{13}C or ^{31}P transitions. Indeed, most spectrometers can only be set up to observe one type of nucleus at a time. In addition, the X approximation aids the diagonalization of the Hamiltonian.

The AB Hamiltonian discussed in Section 2-7 was simplified for an AX system [in which $|J| \ll |\nu_A - \nu_X|$], by omitting the last term to give

$$\hat{\mathscr{H}}_{AX} = \nu_A \hat{I}_{Az} - \nu_X \hat{I}_{Xz} + J\hat{I}_{Az}\hat{I}_{Xz}. \tag{2-87}$$

A similar approximation can be made in any system containing more than one type of nucleus. Coupling between nuclei of different types only contributes terms like

$$J_{AX}\hat{I}_{Az}\hat{I}_{Xz} \tag{2-88}$$

to the Hamiltonian, while coupling between nuclei of the same type gives the complete term

$$J_{AB}[\hat{I}_{Az}\hat{I}_{Bz} + \tfrac{1}{2}(\hat{I}_{A+}\hat{I}_{B-} + \hat{I}_{A-}\hat{I}_{B+})]. \tag{2-89}$$

As the simplified Hamiltonian commutes with $\hat{F}_{Az}$ $(= \sum_{A_j} \hat{I}_{jz})$ and $\hat{F}_{Xz}$, the stationary states have definite values of both m_A and m_X, the total A and X spins. The simple product functions can be classified according to the values of m_A and m_X and can be mixed only if both m_A and m_X are the same. This leads to additional factorization of the Hamiltonian matrix. The intensity of a transition $m \leftarrow n$ depends on

$$|\langle m|\, \gamma_A\hat{F}_{A-} + \gamma_X\hat{F}_{X-}\, |n\rangle|^2 \tag{2-90}$$

As the states $|m\rangle$ and $|n\rangle$ have definite values of m_A and m_X, and as $\hat{F}_{A-}$ reduces m_A by one unit, the selection rules are

$$\begin{array}{lll} \text{either} & \Delta m_A = -1, & \Delta m_X = 0 \ (A \text{ transition}) \\ \text{or} & \Delta m_X = -1, & \Delta m_A = 0 \ (X \text{ transition}) \end{array} \tag{2-91}$$

so that all transitions belong to one type of nucleus. The relative intensities of the A lines depend on

$$|\langle m|\, \hat{F}_{A-}\, |n\rangle|^2 \tag{2-92}$$

The X approximation is not confined to nuclei of different types, but can be applied to any groups of nuclei whose relative chemical shifts much exceed their mutual coupling.

Systems with one X-nucleus

The spectra of such systems are of considerable practical importance, because of the occurrence of small percentages of molecules containing protons and a single ^{13}C or ^{29}Si, which have spin $\frac{1}{2}$. The normal isotopes of both carbon and silicon are non-magnetic. There are, of course, many other molecules with one X nucleus, such as vinyl fluoride and dimethylphosphine. The spectra of such systems can be analysed in terms of sub-spectra of familiar types. This concept can best be illustrated by the ABX system, the AB portion of which consists of two ab sub-spectra (Fig. 2-4).

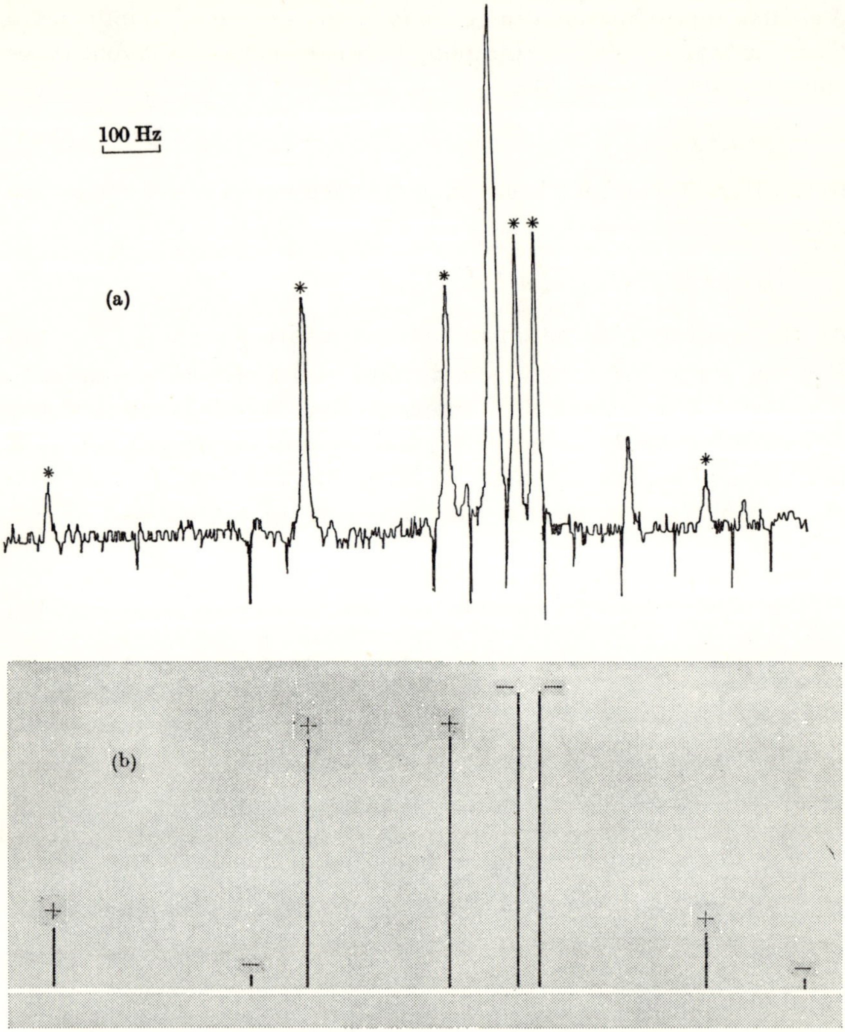

Fig. 2-4 (a) The ^{31}P spectrum of a saturated aqueous solution of sodium diphosphite at 24·29 MHz. The lines of the diphosphite ion, $[HP_2O_5]^{3-}$, are shown by asterisks (the remaining intense line is due to phosphate ion). The spectrum was obtained using a computer to average many sweeps; this procedure increases signal-to-noise ratio. (b) Calculated *AB* part of an *ABX* spectrum to correspond to (a). The breakdown into two *ab* sub-spectra is shown by the plus and minus signs.

The *ABX* spin Hamiltonian is

$$\begin{aligned}\hat{\mathscr{H}}_{ABX} = & -\nu_A \hat{I}_{Az} - \nu_B \hat{I}_{Bz} + J_{AB} \hat{\mathbf{I}}_A \cdot \hat{\mathbf{I}}_B \\ & - \nu_X \hat{I}_{Xz} \\ & + J_{AX} \hat{I}_{Az} \hat{I}_{Xz} + J_{BX} \hat{I}_{Bz} \hat{I}_{Xz}\end{aligned} \qquad (2\text{-}93)$$

This commutes with $\hat{I}_{Xz}$ and $\hat{F}_{ABz}$ $(= \hat{I}_{Az} + \hat{I}_{Bz})$. The part of the spectrum involving transitions of the A and B nuclei has the selection rule

$$\Delta m_{AB} = -1, \qquad \Delta m_X = 0 \tag{2-94}$$

so the transitions can be divided into two sets, those between states in which $m_X = +\frac{1}{2}$ (α transitions) and those for which $m_X = -\frac{1}{2}$ (β transitions). The molecules in which the X spin is the state $|\alpha\rangle$ will have a Hamiltonian

$$\hat{\mathscr{H}}_{AB\alpha} = -(\nu_A - \tfrac{1}{2}J_{AX})\mathbf{I}_{Az} - (\nu_B - \tfrac{1}{2}J_{BX})\hat{I}_{Bz} + J_{AB}\mathbf{I}_A \cdot \mathbf{I}_B - \tfrac{1}{2}\nu_X \tag{2-95}$$

As this is like an AB Hamiltonian with ν_A replaced by $(\nu_A - \frac{1}{2}J_{AX})$, ν_B by $(\nu_B - \frac{1}{2}J_{BX})$ and with a constant term added, the α transitions form an *ab* sub-spectrum which is identical to an AB spectrum with *effective* chemical shifts $\nu_A - \frac{1}{2}J_{AX}$ and $\nu_B - \frac{1}{2}J_{BX}$. Similarly, the β transitions form an *ab* sub-spectrum with chemical shifts $\nu_A + \frac{1}{2}J_{AX}$ and $\nu_B + \frac{1}{2}J_{BX}$. Small letters are used to designate sub-spectral types. The X part of the spectrum cannot be analysed into sub-spectra.

Any system ABC—NX with one X nucleus has two *abc—n* sub-spectra in the non-X regions corresponding the two X spin states. These sub-spectra have the same coupling constants as the original system but effective chemical shifts $\nu_j \pm \frac{1}{2}J_{jX}$. This simplifies the interpretation of such spectra.

Satellite spectra

About 1% of naturally occurring carbon nuclei are the magnetic isotope ^{13}C with spin $\frac{1}{2}$. The proton resonance spectrum of an organic compound comes both from those molecules with non-magnetic carbon and from molecules in which one carbon nucleus is magnetic. Although the lines from the latter are much weaker they can be detected with care. Called ^{13}C satellites, they appear on either side of the normal ^{12}C peaks separated by $J(^{13}CH)$.

Figure 2-5 shows the ^{13}C satellites of 1,1,2,2-tetrabromoethane. The two protons in the normal molecule $(^{12}CHBr_2)_2$ are magnetically equivalent and give a single spectral line. About 2% of the molecules [**1**] are $^{12}CHBr_2{}^{13}CHBr_2$, in which the protons are coupled differently to the ^{13}C nucleus. The proton spectrum consists of two *ab* sub-spectra with chemical shifts $(+\frac{1}{2}\,{}^1J_{CH}, +\frac{1}{2}\,{}^2J_{CH})$ and $(-\frac{1}{2}\,{}^1J_{CH}, -\frac{1}{2}\,{}^2J_{CH})$ relative to the normal molecule and with coupling constants $^3J_{HH}$. As the $^2J_{CH}$ coupling is small, the *b* lines of the sub-spectra are obscured by the peak from the unsubstituted molecule, but the *a* lines can be clearly seen. These satellites, separated by $^1J_{CH}$,

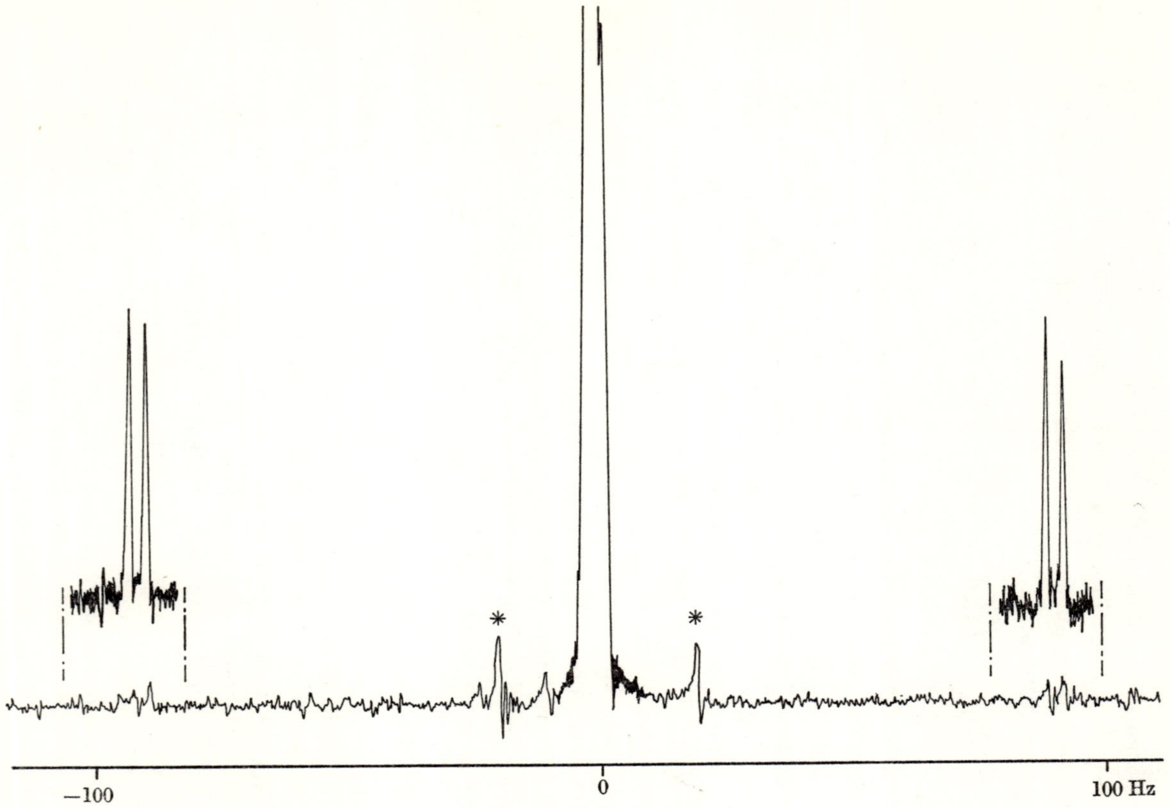

Fig. 2-5 Proton spectrum of 1,1,2,2,-tetrabromoethane showing ^{13}C satellites. These have a doublet structure due to $^{3}J_{HH}$ and are symmetrically placed about the main peak with a separation of $^{1}J_{CH}$ (182 Hz); spinning side bands of the main peak are marked with an asterisk. For an explanation of the J symbolism, see Section 3-14.

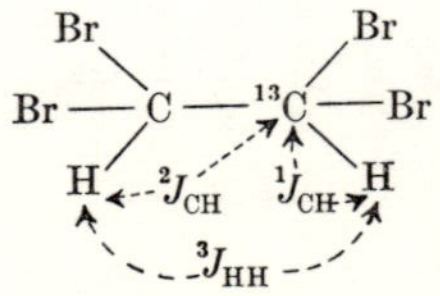

[1]

show a doublet structure due to the *ab* coupling. Isotopic substitution in symmetrical molecules may lift the magnetic equivalence of nuclei and allow normally inaccessible coupling constants to be obtained.

2-10 Equivalent nuclei

In Section 1-15, it was stated that any coupling within a set of magnetically equivalent nuclei does not affect the spectrum. This is most easily understood by considering a set of *completely* equivalent nuclei, which have the same chemical shift and the same coupling to any nucleus in the molecule. In addition to satisfying the criteria for magnetic equivalence, all coupling constants between nuclei of the equivalent set are equal. For example, the three protons of a freely rotating methyl group are *completely* equivalent, as the coupling is the same between any pair of protons, but the six methyl protons of $(CH_3)_2PH$ are only *magnetically* equivalent, as the coupling within the methyl group is different from the coupling between methyl groups.

The Hamiltonian for a molecule containing a set of completely equivalent A nuclei labelled $(1 \ldots j \ldots n)$ and one other nucleus B (an $[A_nB]$ system) is

$$\hat{\mathscr{H}} = -\nu_A \sum_j \hat{I}_{jz} - \nu_B \hat{I}_{Bz} + J_{AA} \sum_{j<k} \hat{\mathbf{I}}_j \cdot \hat{\mathbf{I}}_k + J_{AB} \sum_j \hat{\mathbf{I}}_j \cdot \hat{\mathbf{I}}_B \quad (2\text{-}96)$$

This can be rewritten in terms of the total spin operator of the A nuclei $\hat{\mathbf{F}}_A$ defined as

$$\hat{\mathbf{F}}_A = \sum_j \hat{\mathbf{I}}_j \quad (2\text{-}97)$$

using

$$\hat{\mathbf{F}}_A^2 = \sum_j \hat{\mathbf{I}}_j^2 + 2 \sum_{j<k} \hat{\mathbf{I}}_j \cdot \hat{\mathbf{I}}_k \quad (2\text{-}98)$$

to give

$$\hat{\mathscr{H}} = -\nu_A \hat{F}_{Az} - \nu_B \hat{I}_{Bz} + \tfrac{1}{2} J_{AA}(\hat{\mathbf{F}}_A^2 - \sum_j \hat{\mathbf{I}}_j^2) + J_{AB}\mathbf{F}_A \cdot \hat{\mathbf{I}}_B \quad (2\text{-}99)$$

As $\hat{\mathbf{F}}_A$ is an angular momentum operator it obeys all the usual commutation relations; $\hat{\mathbf{F}}_A^2$ commutes with $\hat{F}_{Az}$, $\hat{F}_{A+}$, and $\hat{F}_{A-}$; it also commutes with $\hat{\mathbf{I}}_B$ as the latter acts on different coordinates. Hence it commutes with $\hat{\mathscr{H}}$ and the stationary states are eigenfunctions of $\hat{\mathbf{F}}_A^2$ with eigenvalues of the form $[A(A+1)]$. The energy can be written

$$E = E'(\nu_A, \nu_B, J_{AB}) + \tfrac{1}{2} J_{AA}[A(A+1) - \tfrac{3}{4}n] \quad (2\text{-}100)$$

The intensity of a transition from state n to state m depends on

$$|\langle m|\ \hat{F}_-\ |n\rangle|^2 = |\langle m|\ \hat{F}_{A-} + \hat{I}_{B-}\ |n\rangle|^2 \tag{2-101}$$

As $\hat{F}_-$ commutes with $\hat{\mathbf{F}}_A^2$, this integral only connects states with the same eigenvalue of $\hat{\mathbf{F}}_A^2$. The frequency of an allowed transition is therefore

$$\nu_{nm} = \hbar^{-1}(E'_m - E'_n) \tag{2-102}$$

which does not depend on J_{AA}. Thus J_{AA} does not affect the spectrum. This proof can easily be extended to a system with more nuclei (e.g. A_nBC).

The proof for magnetically equivalent nuclei follows the same lines. The complete Hamiltonian is written

$$\hat{\mathscr{H}} = \hat{\mathscr{H}}_1(\nu_A, \nu_B, J_{AB}) + \hat{\mathscr{H}}_2(J_{AA}) \tag{2-103}$$

and the two parts are shown to commute; $\hat{F}_+$ commutes with the second term which therefore cannot affect the spectrum (Problem 2-15).

First-order spectra containing groups of magnetically equivalent nuclei show multiplets with equally spaced lines with characteristic intensity patterns. A nucleus X, which is coupled to a set of n magnetically equivalent nuclei A_n, has transitions at frequencies $\nu_X + J_{AX}m_A$, where the values of m_A range from $\frac{1}{2}n$ to $-\frac{1}{2}n$. The relative intensities of these lines depend on the number of A states with each value of m_A. If the A nuclei have spin $\frac{1}{2}$ the relative intensities are the binomial coefficients ^mC_n as shown in Section 1-15.

2-11 The composite particle method for systems containing magnetically equivalent nuclei (Not necessarily 1st order)

As the coupling within a set of magnetically equivalent nuclei does not affect the spectrum, it can be omitted from the calculation. The remaining terms in the spin Hamiltonian involve only the spin operators for the total spin of the group and not the individual spins. The spectrum of the $[A_nB]$ system, for example, can be calculated from the effective Hamiltonian

$$\hat{\mathscr{H}} = -\nu_A\hat{F}_{Az} - \nu_B\hat{I}_{Bz} + J_{AB}\hat{\mathbf{F}}_A \cdot \hat{\mathbf{I}}_B \tag{2-104}$$

As this Hamiltonian does not contain the individual A nuclear spin operators, but only $\hat{\mathbf{F}}_A$, the A_n group can be replaced by a composite particle $\mathscr{A}$ with spin states that are eigenfunctions for $\hat{\mathbf{F}}_A^2$ and $\hat{F}_{Az}$. There is one important difference between an ordinary nucleus and a composite particle, which is that several possible eigenvalues of $\hat{\mathbf{F}}_A^2$

must be considered. Thus the $[A_nB]$ spectrum is a superposition of $\mathscr{A}B$ sub-spectra with different values of the spin of the composite particle $\mathscr{A}$.

The spin states of a composite particle are the states $|F_A, m_A\rangle$. These are used as the basis for setting up the Hamiltonian matrix. Possible values of F are found by coupling the n nuclear spins together in the same way that electronic angular momenta are coupled to find total L and S values in the Russell–Saunders scheme for many-electron atoms (Problem 2-16). As some values of F can be obtained in several different ways, the corresponding composite particle has a degeneracy.

Table 2-4 Composite particle states for n equivalent nuclei

F	$n = 1$	2	3	4	5	6
0	–	S	–	2S	–	5S
$\frac{1}{2}$	D	–	2D	–	5D	–
1		T	–	3T	–	9T
$\frac{3}{2}$			Q	–	4Q	–
2				Qt	–	5Qt
$\frac{5}{2}$					Sx	–
3						Sp

Table 2-4 shows the F values allowed for different numbers of magnetically equivalent spin $\frac{1}{2}$ nuclei and their degeneracies. The letters used for the various F values are derived from singlet (S), doublet (D), triplet (T) etc. A singlet state with $F = 0$ has no spectrum and shows no coupling to other nuclei. It should also be noted that the D_AT_B sub-spectra of $[AB_2]$, $[AB_4]$, $[AB_6]$ etc. systems are identical, but their relative intensities differ, as the calculated intensity of each line in a given sub-spectrum must be multiplied by the degeneracy of the corresponding composite particle to give the correct total intensity.

The $[A_3B]$ system

To demonstrate the application of this method we shall calculate the form of the $[A_3B]$ spectrum. The $[A_3]$ group can be replaced by composite particles D_A with weight 2, and Q_A (non-degenerate). The D_AD_B sub-spectrum is an *ab* spectrum, as a composite particle of spin $\frac{1}{2}$ behaves in the same way as a nucleus of spin $\frac{1}{2}$. The Q_AD_B sub-spectrum is similar to the spectrum of two nuclei with spin $\frac{3}{2}$ and $\frac{1}{2}$. Table 2-5 shows the matrix elements of the spin Hamiltonian. The off-diagonal elements come from the term

$$\tfrac{1}{2}J_{AB}(\hat{F}_{A+}\hat{I}_{B-} + \hat{F}_{A-}\hat{I}_{B+}) \qquad (2\text{-}105)$$

Table 2-5 Hamiltonian matrix elements for $[A_3B]$ system

1. Quartet states

Basis state[(a)]	m_T	$\mathscr{H}_{ii}$	$\mathscr{H}_{ij}$
1. $\frac{3}{2}\alpha$	2	$-\frac{3}{2}\nu_A - \frac{1}{2}\nu_B + \frac{3}{4}J$	–
2. $\frac{3}{2}\beta$	1	$-\frac{3}{2}\nu_A + \frac{1}{2}\nu_B - \frac{3}{4}J$	$\mathscr{H}_{23} = \sqrt{3}J/2$
3. $\frac{1}{2}\alpha$	1	$-\frac{1}{2}\nu_A - \frac{1}{2}\nu_B + \frac{1}{4}J$	
4. $\frac{1}{2}\beta$	0	$-\frac{1}{2}\nu_A + \frac{1}{2}\nu_B - \frac{1}{4}J$	$\mathscr{H}_{45} = J$
5. $-\frac{1}{2}\alpha$	0	$\frac{1}{2}\nu_A - \frac{1}{2}\nu_B - \frac{1}{4}J$	
6. $-\frac{1}{2}\beta$	-1	$\frac{1}{2}\nu_A + \frac{1}{2}\nu_B + \frac{1}{4}J$	$\mathscr{H}_{67} = \sqrt{3}J/2$
7. $-\frac{3}{2}\alpha$	-1	$\frac{3}{2}\nu_A - \frac{1}{2}\nu_B - \frac{3}{4}J$	
8. $-\frac{3}{2}\beta$	-2	$\frac{3}{2}\nu_A + \frac{1}{2}\nu_B + \frac{3}{4}J$	–

2. Doublet states

1. $\alpha\alpha$	1	$-\frac{1}{2}\nu_A - \frac{1}{2}\nu_B + \frac{1}{4}J$	–
2. $\alpha\beta$	0	$-\frac{1}{2}\nu_A + \frac{1}{2}\nu_B - \frac{1}{4}J$	$\mathscr{H}_{23} = \frac{1}{2}J$
3. $\beta\alpha$	0	$+\frac{1}{2}\nu_A - \frac{1}{2}\nu_B - \frac{1}{4}J$	
4. $\beta\beta$	-1	$\frac{1}{2}\nu_A + \frac{1}{2}\nu_B + \frac{1}{4}J$	–

(a) The four Q states $|\frac{3}{2}, \frac{3}{2}\rangle$, $|\frac{3}{2}, \frac{1}{2}\rangle$, $|\frac{3}{2}, -\frac{1}{2}\rangle$, $|\frac{3}{2}, -\frac{3}{2}\rangle$ are here denoted by their m values.

F_{A+} raises the value of m_A by one unit, but also gives a factor $[F(F+1) - m(m+1)]^{1/2}$.

The intensities of the allowed transitions are given by $|\langle m|\,\hat{F}_{A-} + \hat{I}_{B-}\,|n\rangle|^2$. Figure 2-6 shows a typical $[A_3B]$ spectrum.

2-12 Analysis of observed spectra

In order to determine chemical shifts and coupling constants from an observed spectrum, the first step is to determine the type of spectrum (ABC, AX etc.). This is normally known from the molecular structure. If the system is simple (e.g. AB) then the line positions can be written explicitly in terms of the ν_i's and J's. Once the lines are assigned, the numerical values of these parameters can be found by arithmetic. If the system is more complex it is less easy to determine the parameters. Some or all of the parameters may be obtainable by a partial analysis, for in most cases some of the line positions can be written explicitly in terms of the ν_i's and J's. These values are then used to calculate the complete spectrum. In other situations the parameters are obtained by a process of iteration. Probable values, chosen by guesswork or from experience, are used to calculate a spectrum. These trial parameters are then varied until observed and calculated spectra agree.

This process is greatly eased by the use of a computer and there are programmes available. Sometimes there is more than one set of

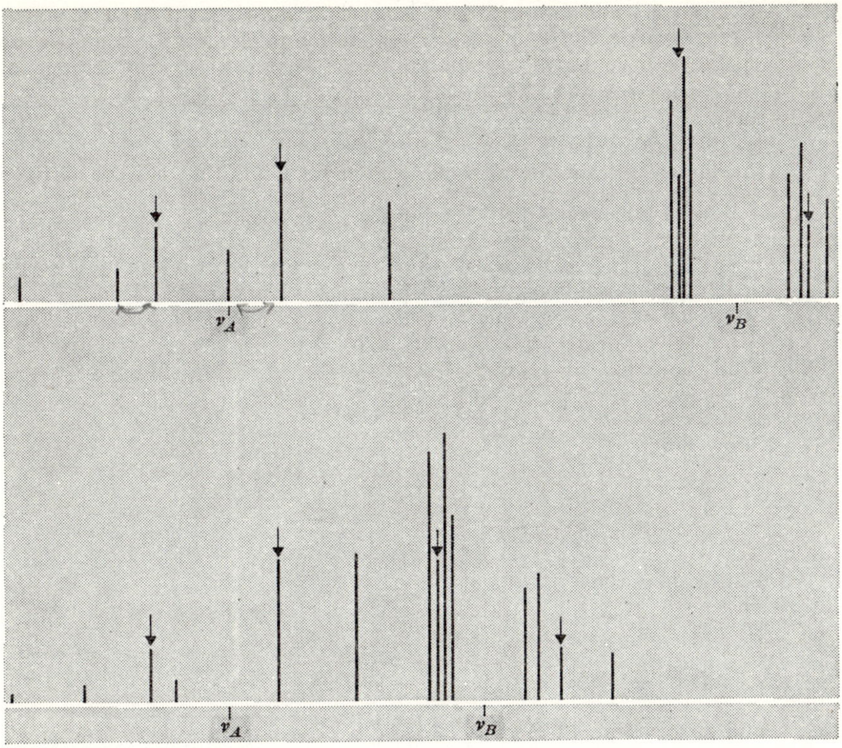

Fig. 2-6 Calculated AB_3 spectra with (a) $\nu_A - \nu_B = 4J$ and (b) $\nu_A - \nu_B = 2J$. The D_AD_B sub-spectral lines which form an *ab* sub-spectrum are indicated by arrows. These spectra show the increasing deviation from the AX_3 pattern (A quartet and B doublet) as $J/(\nu_A - \nu_B)$ increases.

parameters which fit the line positions of the observed spectrum. Often these have different relative signs for the coupling constants. It is possible in favourable cases to distinguish between these by comparing the observed and calculated intensity patterns.

Problems

2-1 Show that the functions $\exp m\phi$ are eigenfunctions of the operators $\dfrac{\hat{d}}{d\phi}$ and $\hat{R}(\phi')$ where the latter rotates the coordinate system by an angle $-\phi'$ about the z axis. For what values of m are these functions single valued? The operator $\hat{I}_z$ is given by $-i\dfrac{\hat{d}}{d\phi}$ in spherical polar coordinates. Comment on the allowed values of m.

2-2 Find the values of

$$\left[\hat{x}, \frac{\hat{d}^2}{dx^2}\right]; \quad \left[\hat{y}, \frac{\hat{d}^2}{dx^2}\right]$$

2-3 If $\hat{A}$ is a Hermitian operator (Section 2-3) show that it has real eigenvalues [substitute $f = g =$ an eigenfunction in Eqn. (2-9)] and that the eigenfunctions belonging to different eigenvalues are orthogonal [substitute different eigenfunctions for f and g in Eqn. (2-9)].

2-4 Show

(a) $\langle a \mid b \rangle = \langle b \mid a \rangle^*$

(b) The condition for an operator to be Hermitian can be written

$$\langle a | \hat{A} | b \rangle = \langle b | \hat{A} | a \rangle^*$$

(c) $|n\rangle = \sum_i |i\rangle \langle i \mid n \rangle$ where the functions $|i\rangle$ form a complete set of orthonormal functions.

2-5 If $[\hat{A}, \hat{B}] = \hat{C}$ and $\hat{A}\,|a, b\rangle = a\,|a, b\rangle$, $\hat{B}\,|a, b\rangle = b\,|a, b\rangle$, show that either $\hat{C} = 0$ or $a = b = 0$ and $\hat{C}\,|a, b\rangle = 0$.

2-6 Find the values of

$$[\hat{I}_z^2, \hat{I}_+], \qquad [\hat{I}_z^2, \hat{I}_z], \qquad [\hat{I}_z, \hat{I}_+\hat{I}_-], \qquad [\hat{I}^2, \hat{I}_{z+}]$$

2-7 Show that

$$\hat{\mathbf{I}}^2 = \hat{I}_z^2 + \tfrac{1}{2}(\hat{I}_+\hat{I}_- + \hat{I}_-\hat{I}_+) = \hat{I}_z^2 + \hat{I}_z + \hat{I}_-\hat{I}_+$$
$$= \hat{I}_z^2 - \hat{I}_z + \hat{I}_+\hat{I}_-$$

Hence show that

$$\hat{I}^2\,|\hat{\mathbf{I}}, m_{\max}\rangle = m_{\max}(m_{\max} + 1)\,|I, m_{\max}\rangle$$

$$\hat{I}^2\,|\hat{\mathbf{I}}, m_{\min}\rangle = m_{\min}(m_{\min} - 1)\,|I, m_{\min}\rangle$$

and derive Eqn. (2-37).

2-8 If $\hat{I}_+\,|I, m\rangle = c\,|I, m + 1\rangle$ show that $c = \sqrt{[I(I + 1) - m(m + 1)]}$. (Show $c^*c = \langle I, m|\,\hat{I}_-\hat{I}_+\,|I, m\rangle$ and write this in terms of $\hat{\mathbf{I}}^2$ and $\hat{I}_z$ operators).

2-9 What are the values of the following?

(a) $\langle I, m + 2|\,\hat{I}_+\,|I, m\rangle$
(b) $\langle I, m - 1|\,\hat{I}_y\,|I, m\rangle$
(c) $\langle \tfrac{3}{2}, \tfrac{1}{2}|\,\hat{I}_+\,|\tfrac{3}{2}, -\tfrac{1}{2}\rangle$
(d) $\langle I, m + 1|\,\hat{I}_x\,|I, m\rangle$
(e) $\langle \tfrac{3}{2}, \tfrac{3}{2}|\,\hat{I}_+\,|\tfrac{3}{2}, \tfrac{1}{2}\rangle$
(*f*) $\langle \tfrac{3}{2}, -\tfrac{1}{2}|\,\hat{I}_-\,|\tfrac{3}{2}, \tfrac{1}{2}\rangle$

2-10 Calculate the values of

$$\langle\alpha\beta\alpha|\,\hat{I}_{2+}\hat{I}_{3-}\,|\alpha\alpha\beta\rangle,\ \langle\alpha\beta\alpha|\,\hat{I}_{2+}\hat{I}_{3-}\,|\beta\alpha\beta\rangle$$

$$\langle\alpha\beta|\,\hat{I}_{2z}\,|\alpha\beta\rangle,\ \langle\alpha\alpha|\,\hat{I}_{1z}\,|\alpha\beta\rangle,$$

$$\langle\alpha\alpha\alpha|\,\hat{F}_+\,|c_1\alpha\alpha\beta + c_2\alpha\beta\alpha + c_3\beta\alpha\alpha\rangle$$

2-11 Show that the states $2^{-1/2}(\alpha\beta + \beta\alpha)$ and $2^{-1/2}(\alpha\beta - \beta\alpha)$ are eigenstates of the A_2 Hamiltonian with eigenvalues $\tfrac{1}{4}J$ and $-\tfrac{3}{4}J$. What is the effect of the operator $\hat{\mathbf{F}}^2$ [$= (\hat{\mathbf{I}}_1 + \hat{\mathbf{I}}_2)^2$] on these states and on the state $\alpha\alpha$? These are T and S composite particle states (Section 2-11).

2-12 Write down the electron spin wave functions of the singlet and triplet states of helium, and the nuclear spin states of ortho and para hydrogen. Show that the singlet wave functions are anti-symmetrical with respect to interchange of particle labels.

2-13 Show that the selection rule $\Delta m_T = -1$ for an AX_m system is equivalent to either $\Delta m_A = -1$ or $\Delta m_X = -1$. From Eqn. (2-47) find the relative rates of absorption of energy by A and X transitions in the AX system.

2-14 Construct the Hamiltonian matrix for the ABX system.

2-15 If $\hat{\mathbf{F}}_A = \sum_{i=1}^{n} \hat{\mathbf{I}}_i$ show that

$$[\hat{F}_{Az}, \hat{F}_{A+}] = \hat{F}_{A+}$$
$$[\hat{\mathbf{F}}_A^2, \hat{I}_{i+}] = 2(\hat{I}_{i+}\hat{F}_{Az} - \hat{I}_{iz}F_+)$$
$$[\hat{\mathbf{F}}_A^2, \hat{F}_{A+}] = 0$$

2-16 Construct a table of product functions of 4 nuclei classified according to the values of m_T. Show that the values of F are 2, 1 and 0 and find the degeneracy of each value. Deduce that four electrons in the configuration $1s^1 2s^1 2p^1 3s^1$ form two ^1P, three ^3P and one ^5P levels.

2-17 $\hat{\lambda}$ is an operator which operates on a nuclear spin function such that

$$\hat{\lambda}\,|I, m\rangle = |I, \; -m\rangle.$$

Show that

$$\hat{\lambda}^2 = 1; \; \langle I, m|\hat{\lambda} = \langle I, \; -m|$$
$$\hat{\lambda}\hat{I}_z\hat{\lambda} = -\hat{I}_z$$
$$\hat{\lambda}\hat{I}_+\hat{\lambda} = \hat{I}_-$$
$$\hat{\lambda}\hat{I}_-\hat{\lambda} = \hat{I}_+$$

2-18 For a system of n nuclear spins

$$\hat{\Lambda} = \prod_j \hat{\lambda}_j$$

show that $\hat{\Lambda}\hat{\mathscr{H}}\hat{\Lambda} = -\nu_j\hat{I}_{zj} + \sum_{j<k} J_{jk}\hat{\mathbf{I}}_j \,.\, \hat{\mathbf{I}}_k$

If $|i\rangle$ are the set of stationary states of $\hat{\mathscr{H}}$, with energies E_i find the stationary states of $\hat{\mathscr{H}}' = -\hat{\Lambda}\hat{\mathscr{H}}\hat{\Lambda}$ and their energies.

Show that systems with Hamiltonians equal to $\hat{\mathscr{H}}$ and $\hat{\mathscr{H}}'$ have identical spectra and deduce that the absolute signs of coupling constants cannot be determined from the NMR spectrum . [$\hat{\mathscr{H}}'$ has stationary states $\Lambda\,|i\rangle$ with energies $-E_i$; show that $\langle i|\, \Lambda\hat{I}_-\Lambda\,|j\rangle = \langle j|\, \hat{I}_-\,|i\rangle$].

3 Chemical shifts and coupling constants

3-1 Introduction

The values of chemical shifts and coupling constants are often used in an entirely empirical way for structure determination. The organic chemist makes frequent use of such facts as (a) methyl protons of $CH_3—CH_2—C$ groups give rise to resonance near δ 1·0, or (b) proton–proton coupling constants for *vicinal* protons *trans* across C═C are greater than those for corresponding *cis* protons. However, this chapter will not be concerned with such applications, but rather will discuss the theoretical and physical attempts to understand the nature of the nuclear spin parameters. It will be assumed that the student will become familiar with the correlative use of chemical shifts and coupling constants through courses and practical work in organic chemistry; the book by Jackman given in the reading list deals with this aspect of NMR.

The energies involved in NMR are extremely small, but the spin parameters may be measured quite accurately. The nuclei may be regarded as probes in order to obtain detailed information about molecules. Both chemical shifts and coupling constants depend on the distribution of electron density near the nuclei and consequently they provide data on bond hybridization, bond dipole moments and susceptibilities, and even on electronic energies. The way in which electron density is affected by substituents and by interactions with solvent can also be investigated. A considerable mass of information has been accumulated in this way, and, whereas it is not yet possible to calculate chemical shifts and coupling constants *ab initio* for any but the simplest molecules, many of the factors affecting these parameters are well known and predictions for new systems are often possible. Most of the discussion of this chapter will be about data for protons, primarily because most experimental NMR studies are for proton resonance. We shall confine ourselves to the parameters as averaged over molecular tumbling in the liquid or solution phase. This is the normal experimental situation, but it is important to remember that both chemical shifts and coupling constants are in principle anisotropic quantities. Theoretically, the averaging process is incorporated by calculating σ or J for three Cartesian directions, adding these three contributions together, and dividing by 3, so that factors of $\frac{1}{3}$ appear in many equations, e.g. Eqns. (3-14) and (3-28). Moreover, certain terms have zero contribution *on average*, which greatly simplifies the situation (see Section 3-12). In special situations, such as when liquid crystal solvents are used, partial orientation of the molecules occurs and the averaging process is not complete. It is usually only in such cases that the absolute signs of coupling constants can be determined; only the relative signs can be determined by measurement of absorption frequencies and intensities when there is no preferred orientation of the molecules.

PART A CHEMICAL SHIFTS

3-2 Atomic screening

As discussed in Section 1-7, an external magnetic field induces motion of electrons such that a secondary magnetic field is produced which opposes the primary field at the centre of the motion (a diamagnetic effect) and therefore *lowers* the frequency needed for resonance. Such motion may be described as an effective rotation of the whole electronic cloud about the direction of $\mathbf{B}_0$. The rotation occurs at an angular velocity $\boldsymbol{\omega}_i$ given by

$$\boldsymbol{\omega}_i = \frac{e}{2m_e}\mathbf{B}_0 \tag{3-1}$$

Now such an electronic motion is related to a current density $\mathbf{j}$ (see Fig. 3-1)

$$\mathbf{j} = -e(\boldsymbol{\omega}_i \times \mathbf{r})\rho_e = -\frac{e^2}{2m_e}(\mathbf{B}_0 \times \mathbf{r})\rho_e \tag{3-2}$$

In this equation $\boldsymbol{\omega}_i \times \mathbf{r}$ is the velocity of the electrons, $\mathbf{r}$ is the distance from the nuclear origin to the element of current density under consideration, and ρ_e is the charge density in units of the electronic charge. Such a current density produces an element $d\mathbf{B}_i$ of magnetic

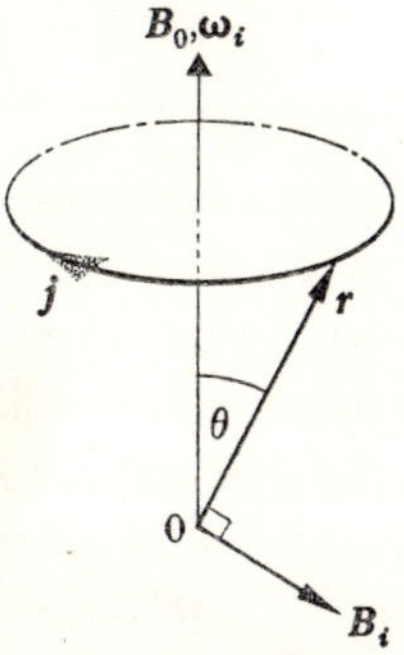

Fig. 3-1 Diagram illustrating diamagnetic circulation of the electronic charge cloud under the influence of a magnetic field $\mathbf{B}_0$.

field at the origin, O, which is given by Biot and Savart's Law (note: we have defined **r** in the opposite direction to that normally used in Biot–Savart calculations) as

$$d\mathbf{B}_i = -\frac{\Xi_0}{4\pi r^3}(\mathbf{j} \times \mathbf{r})\, dV \tag{3-3}$$

where Ξ_0 is the permeability constant, and dV is the element of volume with current density **j**. Substitution for **j** gives

$$d\mathbf{B}_i = \frac{\Xi_0 e^2}{8\pi m_e r^3}(\mathbf{B}_0 \times \mathbf{r}) \times \mathbf{r}\rho_e\, dV \tag{3-4}$$

If **r** is at an angle θ to $\mathbf{B}_0$ (Fig. 3-1) this reduces to

$$dB_i = \frac{\Xi_0 e^2}{8\pi m_e r} B_0 \rho_e \sin\theta\, dV \tag{3-5}$$

This field has a component along $\mathbf{B}_0$ of $-dB_i \sin\theta$ (components perpendicular to B_0 are zero on average). Integration over all space gives the total component of the induced field

$$B_{iz} = -\frac{\Xi_0 e^2 B_0}{8\pi m_e}\int\left(\frac{\rho_e}{r}\right)\sin^2\theta\, dV \tag{3-6}$$

The diamagnetic contribution, σ_d, to the shielding constant is therefore (see Eqn. 1-22)

$$\sigma_d = \frac{\Xi_0 e^2}{8\pi m_e}\int\left(\frac{\rho_e}{r}\right)\sin^2\theta\, dV \tag{3-7}$$

Transformation to polar coordinates, followed by integration (Problem 3-1) under the assumption of a spherical charge distribution (valid for atoms) gives

$$\sigma_d = \frac{\Xi_0 e^2}{3m_e}\int_0^\infty r\rho_e\, dr \tag{3-8}$$

This result is known as Lamb's formula; an equivalent expression may be derived quantum mechanically. It may be evaluated if the electron distribution is known. For atoms it gives the *total* shielding constant. For the hydrogen atom σ is calculated to be 17·8 p.p.m., but the value increases rapidly and monotonically with the nuclear atomic number (since the number of electrons increases). It has been estimated as 261 p.p.m. for an isolated ^{13}C atom, and 464 p.p.m. for ^{19}F. However, in NMR experiments we are only concerned with *differences* in screening constants between atoms of the same species in different molecules. Such differences may be expected to be very small fractions of the total diamagnetic effect.

3-3 Molecular screening: the paramagnetic term

In molecules the presence of the other nuclei hinders rotation of the electron cloud about the nucleus whose shielding is being considered (indeed, most electrons will probably be primarily associated with other nuclei and bonds). This statement, though oversimplified, indicates that in molecular systems there exists, in addition to the Lamb term, another contribution to shielding which normally reduces the diamagnetic effect and is therefore known as the paramagnetic* term.

The paramagnetic contribution to shielding may be estimated by finding the way in which the electronic wave functions are modified by the magnetic field. When a field is applied, new energy terms appear, and the electronic wave functions calculated in the absence of the field are no longer eigenfunctions of the electronic Hamiltonian. The new electronic wave functions may be obtained as linear combinations of the eigenfunctions for the field-absent case. Thus the field can be considered to mix 'excited' electronic states into the ground state wave function. The principal extra energy term which appears when the field is applied is the Zeeman energy of interaction of the electron orbital magnetic moment, $\boldsymbol{\mu}$, with the field, and it is given classically by

$$U = -\boldsymbol{\mu} \cdot \mathbf{B} \tag{3-9}$$

This energy is very much smaller than the total electronic energy and so the way in which it modifies the wave functions may be derived using perturbation theory (see the book list for further reading on this topic). The resulting expression for the paramagnetic contribution to shielding (σ_p) involves a summation over 'excited' electronic states, with the energies of excitation in the denominator. The full equation, due to Ramsey, is rather complex and will not be given here; there is an analogous term in the Van Vleck formula for magnetic susceptibility.

The shielding constant for molecules may be expressed as the sum of diamagnetic and paramagnetic parts (the former analogous to the Lamb term for atoms).

$$\sigma = \sigma_d + \sigma_p \tag{3-10}$$

* The expression 'paramagnetic' is somewhat misleading—it should be emphasized that this term (sometimes called the temperature-independent paramagnetic term) is to be sharply distinguished from the temperature-dependent paramagnetism associated with systems having unpaired electrons.

Evaluation of σ requires detailed knowledge of excited state wave functions, which is rarely available. Nonetheless, accurate calculations have been made for a few simple molecules, such as H_2 and LiH. Moreover, some general comments are useful. The paramagnetic term vanishes for *s* electrons, since these are spherically symmetrical with zero orbital angular momentum, and they therefore have no magnetic moment to contribute as in Eqn. (3-9). On the other hand, σ_p may be large (a) for an asymmetric distribution of *p* and *d* electrons near the nucleus in question, and (b) if there are low-lying excited states. Clearly, both σ_p and σ_d will differ for different nuclei in the same molecule, owing to the uneven distribution of electron density. They also depend on the orientation of the molecule relative to the magnetic field, i.e. σ is an anisotropic property, but we will only consider the average obtained due to random molecular tumbling.

3-4 Screening constants for large molecules

For large molecules σ_d and σ_p tend to become very large but of opposite sign, so that if they are calculated as indicated in Sections 3-2 and 3-3 there is a high uncertainty in the net σ. Exact calculations of σ are no longer feasible. However it proves possible to make use of semi-empirical methods. The total screening constant is written as a sum of local effects and long-range or molecular effects. These are further broken down into separate terms as follows

$$\sigma = \sigma_d(\text{local}) + \sigma_p(\text{local}) + \sigma_m + \sigma_r + \sigma_e + \sigma_s \tag{3-11}$$

where σ_m = neighbour anisotropy effect
σ_r = effects of ring currents
σ_e = electric field effects
σ_s = solvent (or medium) effects

This scheme is arbitrary but useful. The effects of σ_d (local) and σ_p(local) are similar to the diamagnetic and paramagnetic terms already discussed, except that only electrons closely associated with the nucleus whose shielding constant is being estimated are considered. Examples of the variations in these terms will now be considered and the remaining contributions to σ will be discussed later. It is pertinent to note at this point that σ_m is a *direct* effect on the magnetic field, whereas σ_e is an *indirect* effect (in that it acts to alter the local terms). The term σ_s contains in principle contributions of the same type as σ_m, σ_r, and σ_e but they are of *inter-* rather than *intra*-molecular origin.

3-5 The local diamagnetic term

Clearly, if the electron density around the magnetic nucleus is reduced, the shielding due to the local diamagnetic term is decreased. This is important for ^1H resonance, as is shown by the following δ values for methyl protons

$Si(CH_3)_4$	CH_3CH_3	MeI	MeCl	Me_2O
0·00	0·88	2·16	3·05	3·24

It has, in fact, been proposed that the chemical shift difference between CH_3 and CH_2 proton resonances in compounds of the type CH_3CH_2X be used to predict a substituent electronegativity, E_X, for X. The inductive effect of X affects the CH_3 resonance less than the CH_2 resonance because of the attenuation due to the extra intervening carbon atom. The relationship (3-12) has been suggested.

$$E_X = 0{\cdot}684(\delta_{CH_2} - \delta_{CH_3}) + 1{\cdot}78 \qquad (3\text{-}12)$$

However, such equations need to be used with caution in view of the other contributions to σ listed in Eqn. (3-11).

3-6 The local paramagnetic term

The local paramagnetic effect depends on the excitation energies of electrons primarily associated with the nucleus under consideration. It produces an induced magnetic moment which reinforces the applied field and therefore leads to *deshielding*, i.e. a high frequency shift in NMR. The chemical shifts of nuclei other than ^1H are often dominated by σ_p(local). For instance the ^{19}F resonance of F_2 occurs 625 p.p.m. to high frequency of that for HF. The difference in σ_d is estimated to be about 23 p.p.m. and therefore cannot account for the observed values. However, σ_p is zero for a spherically symmetric species such as an atom or an ion, e.g. F^-, whereas it is appreciable for an asymmetric distribution of p electrons as for covalently-bonded fluorine. There is therefore a strong dependence of σ_p on the ionic character of the bond to fluorine. In the case of HF the paramagnetic term is very small, while for F_2 it has been estimated at $\sim$2,000 p.p.m.; the calculated effect is more than adequate to account for the observed shift between the two species.

Paramagnetic effects on shielding are particularly marked for transition metal complexes because of the rather small electronic excitation energies. These are described by ligand field theory. Thus Table 3-1 lists the ^{59}Co chemical shifts, in p.p.m. relative to that of $K_3Co(CN)_6$, for a series of cobalt complexes; in general, the shifts correlate with the energy separation, ΔE, between the ground and excited state of the d^6 configuration. Thus the shielding decreases

Table 3-1 Calculated and observed ^{59}Co chemical shifts[a] for a series of cobalt complexes

Compound[b]	*Observed shift*	*Calculated shift*	$h^{-1}c^{-1}\,\Delta E$ (cm^{-1})
$Co(NH_3)_3(NO_2)_3$	−6,940	−6,000	23,210
$[Co(en)_3]Cl_3$	−7,010	−7,700	21,400
$Na_3[Co(NO_2)_6]$	−7,350	−8,500	20,670
$[Co(NH_3)_5NO_2]Cl_2$	−7,460	−7,300	21,840
$[Co(NH_3)_6]Cl_3$	−8,080	−8,200	21,000
$[Co(NH_3)_5Cl]Cl_2$	−9,070	−11,100	18,720
$Co(acac)_3$	−12,300	−14,000	16,900

(a) Negative shifts indicate decreased shielding.
(b) en = ethylenediamine; acac = acetylacetonate ion.

as the ligand-field strength of the ligands decreases, i.e. $|\sigma_p|$ is in the order acac $>$ Cl^- $>$ NH_3 $>$ en $>$ NO_2^- $>$ CN. Values of ΔE and calculated chemical shifts are given in Table 3-1.

It may be asked at this stage why ^1H shifts have a much smaller range (ca. 20 p.p.m., with a few exceptions) than for other nuclei (e.g. ca. 800 p.p.m. for ^{19}F). There is a twofold answer: in the first place the total number of electrons in the vicinity of ^1H is smaller than for other nuclei so the magnitude of σ is smaller and consequently changes in σ from one molecule to another are also small. Secondly, the lowest excitation energy involved in σ_p for ^1H is of the type $1s - 2p$, whereas for ^{19}F there are lower-lying orbitals; σ_p is thus larger for ^{19}F. Due to the dependence of σ_p on $1/\Delta E$, it is *not* the dominant effect for ^1H chemical shifts.

3-7 Magnetic anisotropy of neighbouring bonds

The magnetic field B_0 will induce magnetic moments in all the electrons of the molecule. Apart from those which are delocalized (Section 3-8), each electron will be associated with an atom or a bond. It is of importance to ask whether induced moments in a group G of bonding or atomic electrons will affect a nucleus X at a distance. Clearly there is such an effect in any given molecular orientation (Fig. 3-2). However, we are usually concerned with molecules that are rapidly tumbling, and it can be shown that provided the induced moment in G is constant (isotropic) the effect at X averages to zero. Induced moments are proportional to magnetic susceptibilities; thus, when the magnetic field intensity is along a principal axis (ℓ) of magnetic susceptibility for G, the induced magnetic moment is

$$\mu_{i\ell}(G) = \chi_\ell(G)H_\ell \qquad (3\text{-}13)$$

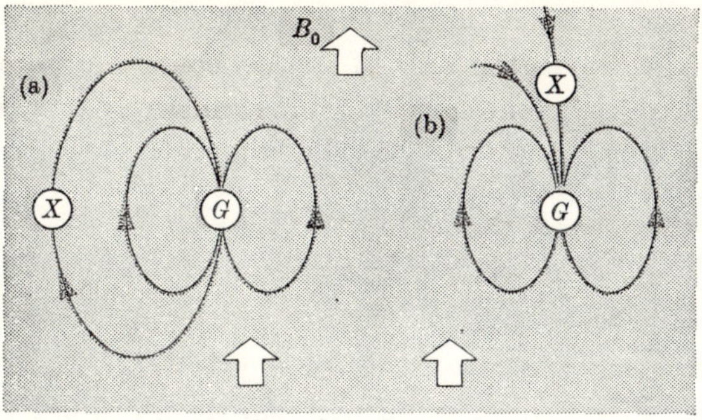

Fig. 3-2 The induced magnetic field produced from an electronic group G, and its effect at a distant nucleus X: (a) $r_{G-X} \perp B_0$ (X deshielded); (b) $r_{G-X} \parallel B_0$ (X shielded).

where χ_ℓ is the susceptibility (per molecule of G in direction ℓ), and H_ℓ is the field intensity. It is clear that only anisotropy in bond susceptibilities causes screening effects at a distance. In fact, it can be shown that if the induced magnetic moments are considered to act as point magnetic dipoles, the contribution to $\Delta\sigma$ from a group G is

$$\Delta\sigma = \frac{1}{3r^3} \sum_{\ell=x,y,z} \chi_\ell(1 - 3\cos^2\theta_\ell)/4\pi \qquad (3\text{-}14)$$

where r is the distance between G and the nucleus X, and θ_ℓ is the angle between axis ℓ and r. In most cases the principal axes for bond susceptibilities will be along or perpendicular to the bond; usually the point dipole induced is considered to be at the bond centre. Only bonding electrons and lone pairs need to be considered, since inner-shell electrons associated with individual atoms will have isotropic susceptibility contributions. Equation (3-14) can be shown to reduce to

$$\Delta\sigma = \frac{1}{3r^3}[(\chi_\parallel - \chi'_\perp)(3\cos^2\theta'_\perp - 1) + (\chi_\parallel - \chi''_\perp)(3\cos^2\theta''_\perp - 1)]/4\pi \qquad (3\text{-}15)$$

where $\chi_\parallel$ is the longitudinal susceptibility (i.e. along the bond direction), $\chi'_\perp$ and $\chi''_\perp$ are the two transverse principal susceptibilities (i.e. perpendicular to the bond), while $\theta'_\perp$ and $\theta''_\perp$ are the angles between r and the appropriate bond perpendicular. The general nature of the shielding is such that there is a nodal cone (of elliptical cross-section) whose axis is in the direction of one of the principle susceptibilities (see Fig. 3-3). Further simplification occurs if it is considered that the bond is cylindrically symmetrical ($\chi'_\perp = \chi''_\perp$)—in such a case the cross-section of the nodal cone is circular, all directions perpendicular to the axis of the cone are equally shielded and the half-angle of the cone is 54°44′. This will apply to a good approximation to all single and triple bonds, when the cone axis is the bond direction. In such a case

$$\Delta\sigma = \frac{1}{3r^3}(\chi_{\parallel} - \chi_{\perp})(1 - 3\cos^2\theta)/4\pi \qquad (3\text{-}16)$$

where θ is the angle between the bond axis and r.

Table 3-2 gives some published values for bond susceptibility

Table 3-2 Bond magnetic susceptibility anisotropies[a] and associated shielding effects

	C—H	C—C	C=C[b]	C=O[b]	N=O[b,c]	C≡C
$\Delta\chi^{(d)}_{\parallel\perp'}$	90	140	150	320	370	−340
$\Delta\chi^{(d)}_{\parallel\perp''}$	90	140	150	420	1,300	−340
$\Delta\sigma^{(e)}_{\parallel}$	−15	−22	−24	−58	−130	54
$\Delta\sigma^{(e)}_{\perp'}$	7·5	11	12	17	−44	−27
$\Delta\sigma^{(e)}_{\perp''}$	7·5	11	12	41	175	−27

(a) The values given are adapted from the research literature and are subject to controversy.
(b) In the cases involving double bonds the $\perp'$ direction is taken to be in the nodal plane of the π electrons.
(c) In N-nitroso compounds.
(d) $\Delta\chi_{\parallel\perp} = \chi_{\parallel} - \chi_{\perp}$; $\Delta\chi$ is in units of 10^{-36} m^3 per molecule (alternatively $\Delta\chi$ may be expressed in m^3 per mole). It should be noted that values of χ in SI units are 4π times those in e.m.u., due to rationalization.
(e) $\Delta\sigma$ is in units of $10^{-36}/3r^3$, with r in m; $\Delta\sigma_j$ refers to shielding in the direction of axis j.

anisotropies and Fig. 3-3 shows some of the shielding patterns diagrammatically. It may be seen that except for C≡C all the $\Delta\chi$ are positive and there is deshielding along the bond direction; there is shielding in the $\perp''$ direction for all groups involving double bounds. It is assumed that single and triple bonds have cylindrical symmetry; the shielding effect is almost cylindrically symmetrical for C=C, but is markedly not so for C=O and N=O. In fact for N=O there appears to be deshielding in the $\perp'$ direction, and the nodal cone for shielding, in contrast to the other cases, does not have its axis along the bond direction. The magnitude of the effect is large for the N=O group and accounts for the fact that an observable chemical shift dependence on stereochemistry is found for protons β and even γ to N-nitroso groups when shift contributions from other

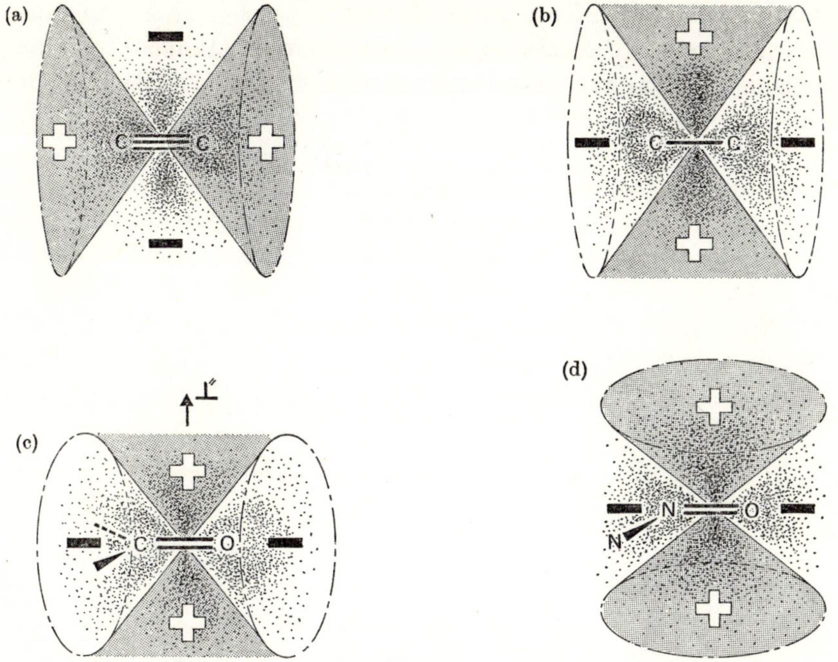

Fig. 3-3 The nodal cones of shielding arising from anisotropy of bond magnetic susceptibility for (a) the C≡C triple bond, (b) the C—C single bond, (c) the carbonyl group, and (d) the nitroso group (in *N*-nitrosamines). The effective centres of the bond magnetic dipoles are not necessarily midway along the bonds. The + and − signs refer to the effect on the shielding constant, σ, of a distant nucleus in the regions of space designated. The magnitude of the effect decreases rapidly as the distance from the bond increases.

sources are probably zero due to local symmetry, e.g. in [**1**]. In less symmetrical situations, it is more difficult to evaluate the contribution of magnetic susceptibility anisotropy. However, the important difference in shielding for axial and equatorial protons of cyclohexane (0·462 p.p.m., with the equatorial proton resonance to high frequency) is attributed to the combined effects of magnetic anisotropy of the C—C and C—H bonds.

The anisotropy of bond susceptibilities can be separated into anisotropy of the diamagnetic term and anisotropy of the temperature-independent paramagnetic term. A useful physical picture can be

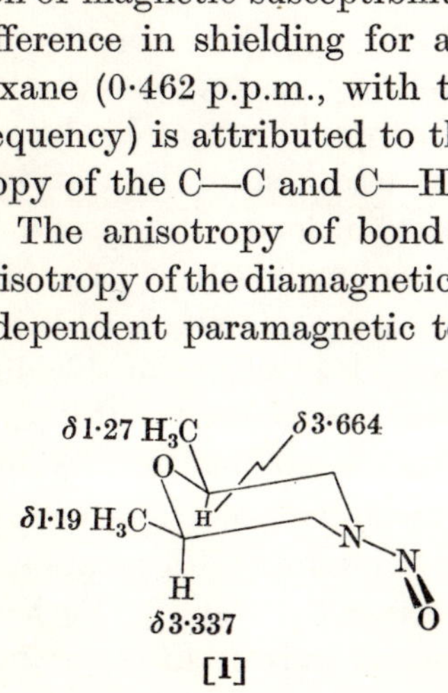

[1]

given in some cases, for instance that for acetylene, where the anisotropy is chiefly in the paramagnetic contribution. When B_0 is along the molecular axis there is no hindrance to circulation of the electrons in the C≡C bond; there is thus a zero paramagnetic effect (Fig. 3-4). When B_0 is perpendicular to the axis the nuclei hinder the circulation, and the paramagnetic effect is strong. An alternative (and more rigorous) picture is given by considerations of the mixing of excited states into the ground state by the magnetic field: mixing

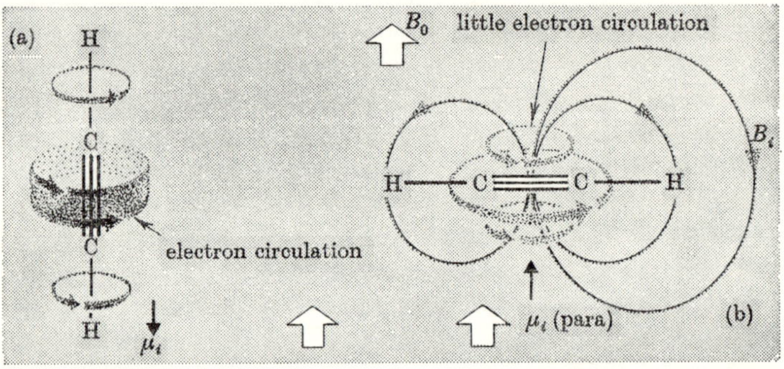

Fig. 3-4 Magnetic anisotropy effects for acetylene: (a) free electron circulation when B_0 is parallel to the C≡C bond, (b) hindered electron circulation when B_0 is perpendicular to the C≡C bond, giving a paramagnetic moment and shielding of the protons.

is possible for B_0 perpendicular to the molecular axis but not for B_0 parallel to the axis. Tumbling averages the paramagnetic term, but as far as the effect of the C≡C electrons on the proton resonance is concerned orientation (b) provides the dominant contribution, and the net result is one of shielding. Clearly this conclusion relies on the anisotropy of the susceptibility of the C≡C bond (see Table 3-2) and is in accord with Fig. 3-3.

This effect explains one apparent oddity in proton chemical shifts—for the series C_2H_2, C_2H_4, C_2H_6. One would expect δ values to decrease in that order due to the inductive effect, since the acetylene protons are most acidic and should be least shielded. However, it is found that in the gas phase the order of shielding is C_2H_4, C_2H_2, C_2H_6, with the shift between C_2H_2 and C_2H_6 (0·6 p.p.m.) much less than that between C_2H_2 and C_2H_4 (3·8 p.p.m.). However, the shielding pattern due to magnetic anisotropy is very different for the three molecules—even the sign of the effect is different for C_2H_2 as compared to C_2H_4 and C_2H_6 (Fig. 3-3). The difference arises because for B_0 parallel to the C—C bond there is hindrance to rotation of the electron cloud for C_2H_4 and C_2H_6 [contrast Fig. 3-4(a)]. Furthermore the protons in C_2H_2 are on a principal axis of susceptibility, which is not the case for C_2H_4 and C_2H_6. In accordance with Eqn. (3-16) there is a large low-frequency shift for the acetylene protons.

It may be noticed that although the local effect of the paramagnetic term is deshielding, its long-range effect due to anisotropy may be shielding (and vice versa for the diamagnetic term). An extreme example is provided by the proton resonance of metal hydrides; δ values as low as -40 have been reported. Some representative δ values are as follows

trans-$IrHCl_2(PEt_3)_3$	$-12{\cdot}6$	$ReH_7(PEt_2PH)_2$	$-5{\cdot}8$
cis-$IrHCl_2(PEt_3)_3$	$-21{\cdot}6$	$OsH_2(CO)_4$	$-8{\cdot}7$
trans-$PtHCl(PEt_3)_2$	$-16{\cdot}8$	$[RhH(NH_3)_5]SO_4$	$-17{\cdot}1$
trans-$PtHNO_3(PEt_3)_2$	$-23{\cdot}6$		

These shifts have been used as diagnostic for metal–hydrogen bonds, since there are few other examples of proton Larmor frequencies to low frequency of TMS. This feature correlates with the high-frequency shifts for resonance of the metal itself, already discussed; both effects arise from the paramagnetic term of the metal. The low-lying *d* electronic levels, distorted from degeneracy by the effect of the ligands, are responsible for the size of the effect.

3-8 Ring currents

When electrons are delocalized, diamagnetic currents may have special importance since the nuclei may not hinder the motion. As the circulations are over large areas, the magnetic moment induced by the field can be quite large. In aromatic systems the delocalized π electrons give rise to a ring current when the field is perpendicular to the molecular plane, as in Fig. 3-5 (though this concept has been criticized). The induced field opposes B_0 at the middle of the molecule but reinforces it at the periphery. Thus the protons in benzene

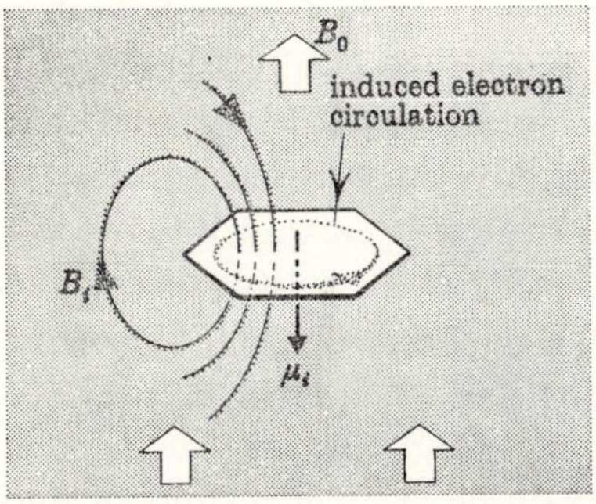

Fig. 3-5 The ring current effect for benzene.

suffer a high-frequency shift—they resonate at δ 7·27 compared to a value of δ ~5·7 calculated for a corresponding ring system with alternating single and double bonds. Note that the geometry of the system is such that the diamagnetic circulation of electrons gives an observed high frequency shift. But, it is possible in certain cases for protons to be situated 'above' the aromatic plane or even 'inside' the ring current. In such circumstances the ring current gives low-frequency shifts. A classic example is provided by [18]-annulene [**2**]. The protons are of two types—the 12 'outer' protons resonate at δ 8·9; the six 'inner' protons give rise to absorption at δ −1·8. In

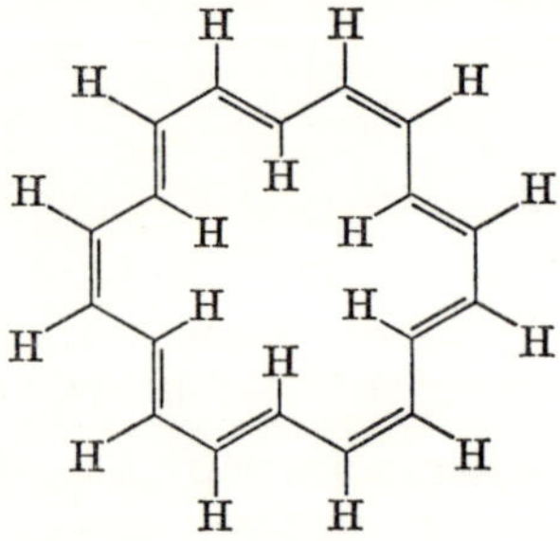

[**2**]

fact the proton spectrum was used as a proof of the aromaticity of [18]-annulene. There have been suggestions that PMR data be used to define ring current, and hence aromaticity, quantitatively. However, such a procedure is very sensitive to the choice of model non-aromatic compounds used for comparison purposes. It has been shown that the 'ring currents' in benzene, thiophen, and furan do not differ significantly and therefore that ring current, resonance energy and aromatic reactivity are not related in any simple way.

The ring current does not occur when B_0 is in the plane of the aromatic molecule. Thus the chemical shift for such systems is in principle highly anisotropic; the observed shift in normal high-resolution work is an average over all orientations. Clearly the molecular susceptibility is also highly anisotropic. Indeed, it is only because of this anisotropy that the ring current leads to observable chemical shift effects at a distance; in a sense shifts due to ring currents are analogous to those arising from anisotropy of bond susceptibility, discussed in the preceding section.

The existence of the ring current effect has been used to study molecular conformations. For example, the ethylenic protons of *trans*- and *cis*-stilbene resonate at δ 6·99 and δ 6·49 respectively. This

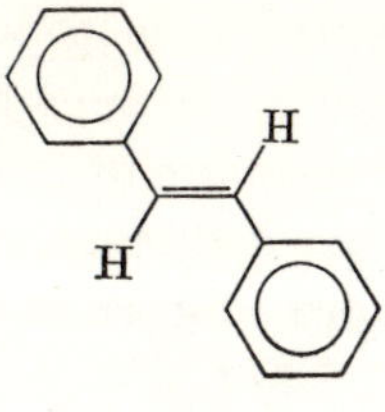

[3]

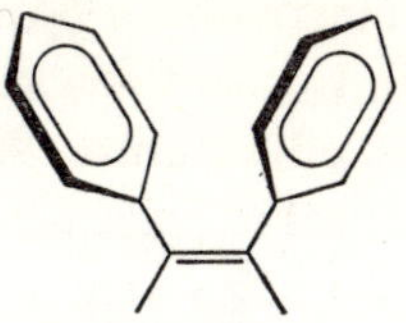

[4]

has been attributed to coplanarity of the aromatic rings (and hence a maximum high-frequency ring current shift) for the *trans* isomer [**3**] but tilting of the rings for the *cis*-isomer [**4**].

3-9 Neighbouring group electric dipoles

If a molecule contains strongly polar groups, there will be intramolecular electric fields. These will have the effect of distorting the electron density in the rest of the molecule, and will therefore affect screening constants. It has been shown that for shielding of a proton in an axially symmetric situation there are two terms contributing to σ_e.

$$\sigma_e = -AE_z - BE^2 \tag{3-17}$$

where z is the direction of the bond to the hydrogen atom, and A and B are constants. The first term may be simply related to drift of electrons along the bond direction (see Fig. 3-6)—the minus sign in Eqn. (3-17) takes account of the fact that electron drift occurs against the direction of **E**. The second term can be viewed as due to an effect on σ_p(local). Consider, for example, an atom: an electric

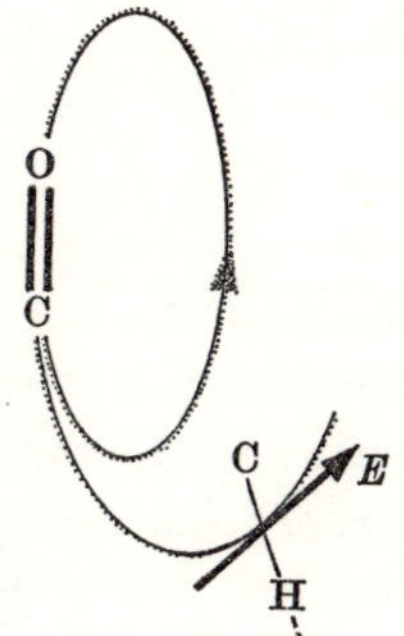

Fig. 3-6 The linear electric field effect on the chemical shift of a C—H proton due to a distant carbonyl group. Note that electrons drift towards the tail of the electric field vector **E**, and therefore, since in this case E_z is negative, the proton is shielded.

field will distort the electronic environment of the nucleus from spherical symmetry and will therefore introduce a paramagnetic shift—the extent of this effect must depend on the square of **E** since the symmetry of the situation requires that a change in direction of **E** should not affect σ. Although at this point we are concerned with intramolecular electric fields, it should be pointed out that similar effects may be obtained in principle from externally applied electric fields. Normally the term in E_z is dominant for proton resonance, and it can either decrease or increase the shielding depending on the sign of E_z. However, the second term always reduces the shielding. In calculations of σ_e it is usual to use a point dipole approximation for the bond electric moments in order to calculate E_z and E^2. For the nitrobenzenes, σ_e may be up to 0·8 p.p.m. Actually, if exact electronic wave functions were known there would be no need to invoke an intramolecular electric field effect, since a full direct calculation of σ_d and σ_p would suffice; however, we need to use Eqn. (3-17) as soon as we start using the approximation of separating local and long-range shielding contributions as discussed in Section 3-4.

The constants A and B differ for different nuclei; in fact for ^{19}F the second term may be more important than the first. Moreover a third term, due to time-dependent electric dipoles mutually induced between a bond X—Y and the bond to the shielded nucleus, is needed. Such dipoles give values of E_z that average to zero, but there is a non-zero average of E^2, written $\langle E^2 \rangle$. This effect has essentially the same origin as van der Waals forces between molecules, and the shielding contribution is therefore known as the van der Waals shift. It has been shown that $\langle E^2 \rangle$ depends on the polarizability and first ionization potential of the electrons in the X—Y bond; there is also an r^{-6} dependence on the distance from the centre of the time-dependent dipoles in the X—Y bond to the shielded nucleus. Chemical shifts between fluorine nuclei in closely related molecules may be dominated by differences in σ_e—for example, the shifts for chloroperfluorobenzenes from the resonance of perfluorobenzene, where the van der Waals contributions are up to 35 p.p.m., and greatly outweigh the E^2 effect of permanent intramolecular electric fields. The differences in this case derive from replacing F by the more polarizable Cl atom. The electric field term accounts for the so-called '*ortho* effect' (the large low-frequency shift associated with the replacement of F by Cl *ortho* to the fluorine nucleus under study). For other series of compounds, a better fit of calculated to observed parameters has been achieved by empirically using different coefficients for the E^2 and $\langle E^2 \rangle$ terms.

3-10 Hydrogen bonds

Some of the highest frequency proton chemical shifts recorded are those of strongly hydrogen-bonded protons, both in intramolecular and in intermolecular systems. For example, the hydroxyl proton in the enol form of acetylacetone [**5**] resonates at $\delta = 15{\cdot}4$. The effect on the proton chemical shift of forming a hydrogen bond cannot be explained on a simple basis; naturally, all the contributions discussed in previous sections must be considered. The electron density in the vicinity of the proton is probably increased by hydrogen-bond formation (though an electrostatic model indicates the reverse)—it is well known that for an O—H · · · O bond the O, O distance is normally less than the van der Waals diameter of oxygen, so the proton lies within the electron cloud of *both* oxygen atoms. Such an increase in electron density should lead to an increase of σ_d(local). Other contributions must therefore predominate in order to bring

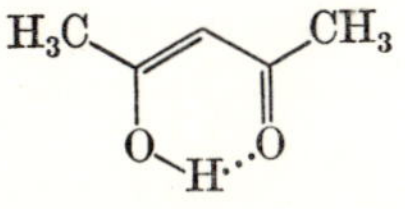

[**5**]

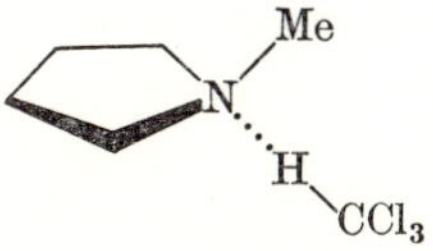

[**6**]

about the substantial deshielding observed. The additional restraint placed on field-induced electronic circulations by the nuclei of the hydrogen bond acceptor may lead to an increase in σ_p(local), giving a high-frequency shift. In addition, effects from susceptibility anisotropy of the acceptor nucleus and of its electric field (including the van der Waals term) are probably substantial.

NMR has been used to measure the strength of hydrogen bonds, but considerable caution has to be used in arriving at quantitative conclusions. High-frequency shifts may, however, be used as criteria for the formation of weak hydrogen bonds. For example, the high-frequency shift for the chloroform proton when in association with *N*-methylpyrrolidine has been attributed to the formation of the weak hydrogen bond complex [**6**]. The resonance of the $CHCl_3$ proton of the complex was found by extrapolation to be 2·05 p.p.m. from that of the free chloroform, and the mole-fraction equilibrium constant for complex formation was determined as 2·2.

By observing chemical shift changes between a pure liquid and a dilute solution in an inert solvent such as CCl_4, NMR provides a useful tool for distinguishing between intra- and intermolecular hydrogen bonding. Only small changes would be observed in the case of intramolecular bonding, but changes of ca. 10 p.p.m. may be found for intermolecular bonding. For example the hydroxyl proton resonance for liquid ethanol occurs to high frequency of the CH_2 proton resonance, but for a dilute solution in $CDCl_3$ the hydroxyl resonance is between those of the CH_2 and CH_3 protons (Fig. 3-7).

The situation for hydroxyl proton resonance (and for NH and NH_2 proton resonances) is complicated by the possible occurrence of

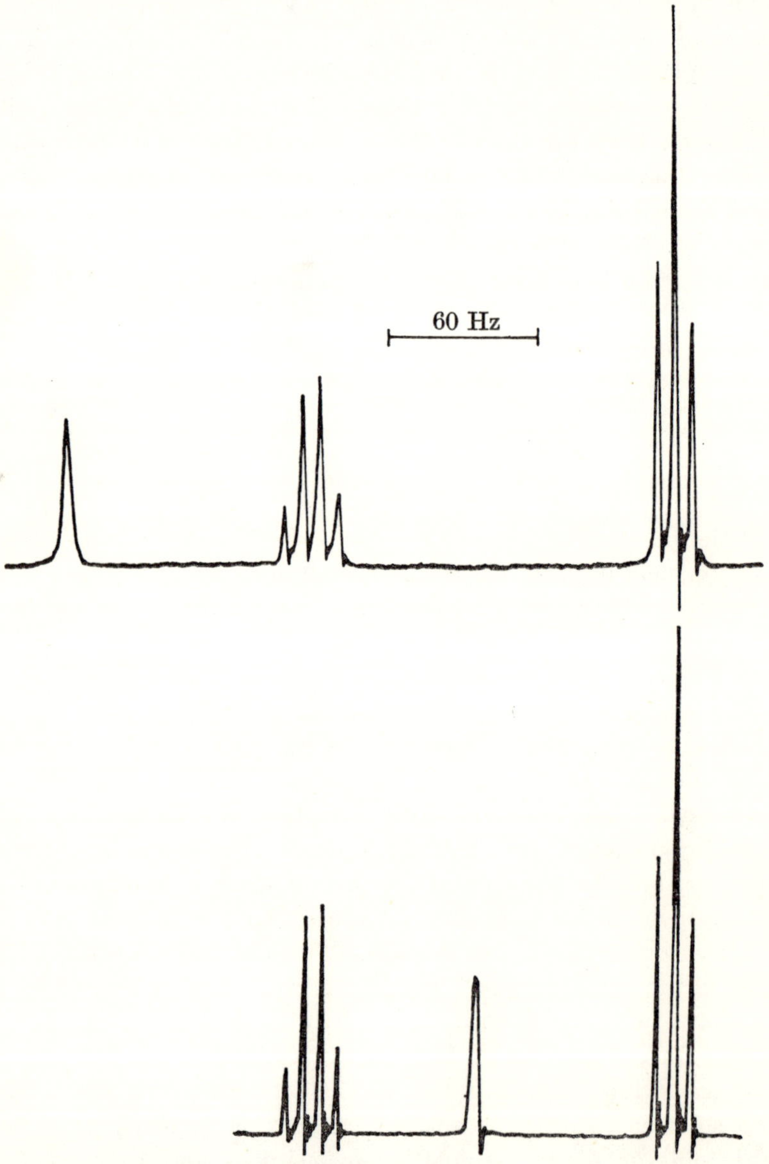

Fig. 3-7 60 MHz PMR spectrum of ethanol: (a) neat liquid, (b) dilute (ca. 7%) solution in $CDCl_3$, to illustrate the effects of intermolecular hydrogen bonding. Note that $CDCl_3$ is not a hydrogen bond acceptor.

non-hydrogen bonded species. In principle there are frequently several different resonances for OH protons over a considerable range of δ values. However, there is usually a rapid exchange of protons between the various environments, and only a single signal is seen at an average position (see Chapter 4). The position of hydroxyl proton resonances will therefore vary greatly because of the variable extent of hydrogen bonding. Since the Larmor frequencies of hydrogen-bonded and non-hydrogen bonded species may in principle be very different, the averaging process may not be sufficiently rapid to give a sharp line. Hydroxyl proton resonances are therefore frequently rather broad.

3-11 Solvent effects

Clearly, solvent molecules can affect resonance positions of a solute nucleus (and vice versa) by the same type of mechanisms as the intramolecular effects already discussed. It is important at the outset to realize that when an internal reference standard such as TMS is used, measured solvent effects actually represent the difference of such effects on the solute and on the standard. Only solvent molecules near the solute or standard will make important contributions. We shall consider solvent effects (which in some situations are better referred to as *medium* effects) under three categories:

(a) The polar solvent effect

This is an electric field effect. A polar solute, such as acetonitrile, will set up a reaction field in a polar solvent (Fig. 3-8). This will occur even for solute molecules with no net dipole moment if they contain polar groups [Fig. 3-8(b)]. The reaction field then causes

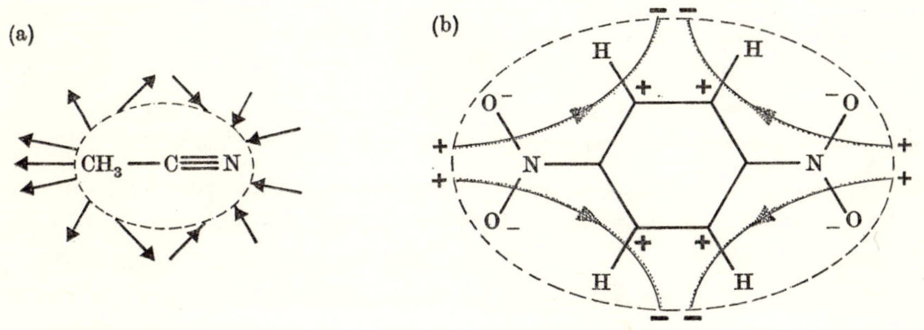

Fig. 3-8 The polar solvent effect: (a) preferred orientation of solvent dipoles around an acetonitrile solute molecule (diagrammatic). The dipoles are drawn with the positive charge at the head of the arrow. The reaction field of the solvent cage acts to augment the electron drift associated with the dipole of the acetonitrile molecule. (b) the reaction field (. . → . .) associated with the quadrupolar molecule *p*-dinitrobenzene.

electron drift in the solute; for polar solutes this occurs so as to augment the shielding or deshielding inductive effect associated with the original electric dipole. This results in changes in shielding. There is also the term in E^2 (Section 3-9) to be taken into account. Thus polar solvents usually tend to augment the inductive effect of electronegative substituents in solutes. In general, protons are at the positive end of molecular or bond dipoles and the action is one of deshielding for solute protons, as in the case of acetonitrile. An isotropic molecule, such as is usually chosen for a standard (e.g. TMS) will be unaffected, except by the van der Waals term.

(b) Solvent magnetic anisotropy

Molecules which have a pronounced anisotropy of magnetic susceptibility will cause shifts when acting as solvents. For example, benzene, due to the ring current effect, causes shielding or deshielding at a distance. Of course, for solutes in benzene solution the effect is averaged over all possible solvent–solute molecular orientations, but because of the disk shape of benzene there is still a net effect, even for spherically symmetrical molecules such as TMS. The effect will not vary with the position in the solute molecule of the nucleus whose shielding is considered, but the magnitude of the shift will differ for different symmetrical solutes. Differential shifts will therefore be observed when TMS is used as a standard, but such effects are probably small.

However, most solutes are not spherically symmetrical, and consequently further more marked differential shifts will occur. Moreover, in this case chemically distinct protons in the same molecule will be affected differently. Note that this may imply that there is a certain preferred orientation between solute and solvent, but it does not necessarily imply any positive attraction between the molecules—there may be simply steric repulsions.

It is frequently possible to make use of such solvent effects to spread out a spectrum and make it amenable to analysis. For this purpose the best solvents are probably benzene (causing general low-frequency shifts relative to TMS due to solute protons coming above or below the benzene plane) or acetone (causing general high-frequency shifts due to solute protons being in the plane of the heavy-atom skeleton of acetone). A good example is δ-1,2,3,4,5,6-hexachlorocyclohexane ([7] gives δ values in benzene solution). Comparison with the chemical shifts in dioxan shows that the resonances of hydrogens A, B, and D have been shifted by ca. 1·0 p.p.m., but the hydrogen C resonance by only ca. 0·25 p.p.m. Clearly, benzene

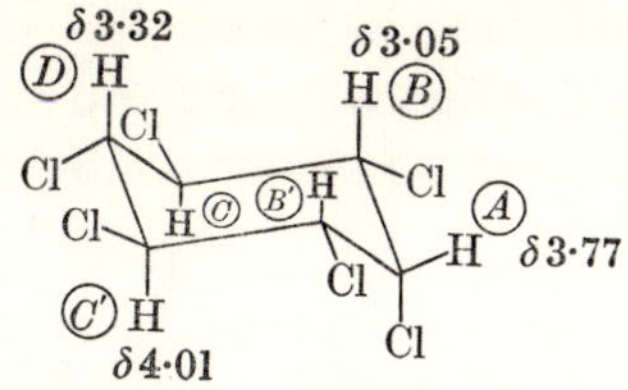

[7]

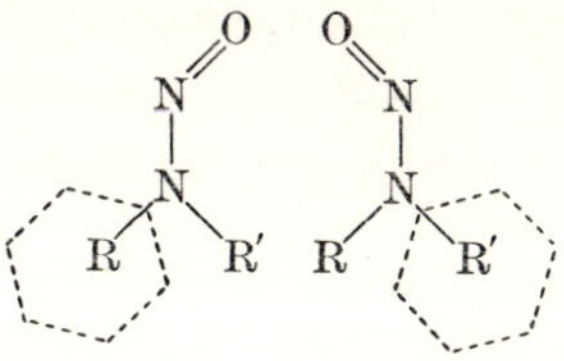

(a) (b)

[8]

molecules can approach closer to one side of [**7**] than to the other (this is reasonably explained on steric grounds by the effects of the chlorine atoms). This results in the unusual feature of the spectrum of the benzene solution: axial protons (C) resonate to high frequency of an equatorial proton (A). The spectrum in benzene, being 'spread out', may be analysed reasonably accurately on a first-order basis.

Such preferred orientation effects (which will be enhanced if there are weak solvent–solute chemical interactions) can frequently lead to useful empirical correlations. For N-nitroso compounds, the benzene ring prefers to lie away from the oxygen atom of the NO group (symbolically the preferred situation is as in [**8a**]), causing greater low-frequency benzene solvent shifts for protons in groups *trans* to the oxygen atom than for those *cis* to oxygen. This effect may be used to distinguish between resonances due to isomers [**8a**] and [**8b**] for $R \neq R'$ in solutions containing mixtures of the two. The magnitude of the effect may be illustrated using the values for dimethylnitrosamine ([**8**], $R = R' = CH_3$). The solvent shift δ(in C_6H_6) $-$ δ(in CCl_4) is $-0{\cdot}56$ p.p.m. for the protons of the *cis* methyl group, and $-0{\cdot}80$ p.p.m. for those of the *trans* group.

(c) Specific interactions

Chemical interaction between solute and solvent, or between two different solutes in the same solution, will clearly cause chemical shift changes. The intermolecularly hydrogen-bonded species discussed in Section 3-10 provide examples of such an effect. NMR evidence is frequently invoked in order to show that chemical interaction occurs. Any type of weak complex formation may be studied by NMR. In fact, there is no clear distinction between 'complex formation' on the one hand and 'preferred orientation' on the other. Measurements of an 'energy of association' become difficult to interpret when these are small (less than about 20 kJ mole^{-1}) and suggestions of a geometry for the complex are unreliable when a whole range of conformations are likely to have similar energies.

However, in many cases it is feasible to talk in terms of a complex, and it becomes possible to obtain chemical shifts of such complexes even when these cannot be isolated. Consider the formation in solution of a charge-transfer complex DA from an electron donor D and an electron acceptor A. The lifetime of an individual complex is normally so small that the NMR signal seen for a nucleus in A, for example, is the average (see Chapter 4) of those for free A and for complex DA, weighted by the respective concentrations. The chemical shift for the solution may be measured relative to that for a similar solution containing no D (i.e. that for free A). Let this relative shift be Δ. Then consideration of the weighted average shows that Eqn. (3-18) applies.

$$\Delta = \Delta_0[A]/([A] + [DA]) \tag{3-18}$$

where the square brackets indicate concentrations and Δ_0 is the chemical shift of A in the pure complex (under solution conditions) relative to that for free A. The equilibrium constant, K_c, for the charge-transfer process may be written

$$K_c = [DA]/[D][A] \tag{3-19}$$

Thence it may be shown (Problem 3-4) that if D is in large excess over A for all the measurements, then

$$\Delta/[D]_0 = -\Delta K_c + \Delta_0 K_c \tag{3-20}$$

where $[D]_0$ is the *total* concentration (i.e. free plus complexed) of D. Thus the plots of $\Delta/[D]_0$ against Δ should be linear, and will give values of both Δ_0 and K_c, even when the former cannot be directly measured because the complex is weak. Fluorine magnetic resonance studies of the acceptor 1,4-dicyano-2,3,5,6-tetrafluorobenzene with the donor hexamethylbenzene in CCl_4 solution give $K_c = 5{\cdot}2$ kg of solution per mole and $\Delta_0 = 2{\cdot}99$ p.p.m. The maximum value of Δ observed (2·26 p.p.m.) was substantially less than Δ_0. NMR thus provides a powerful way of obtaining information about equilibria with high reaction rates; the values of Δ_0 may be used to speculate on the geometry of non-isolable complexes while those of K_c give information about the donor and acceptor properties of D and A respectively.

The chemical shift effects of chemical interaction will have magnetic anisotropy and electric field contributions, and, in addition, will change σ_d(local) and σ_p(local) because the electron distribution in D will be altered.

PART B COUPLING CONSTANTS

3-12 Direct dipolar coupling

Two bar magnets interact such that the energy of the system depends on their mutual orientation. Since many nuclei possess magnetic moments it might be expected that an analogous dipolar interaction would give rise to spin–spin coupling. The classical interaction energy between two magnetic point dipoles $\boldsymbol{\mu}_1$ and $\boldsymbol{\mu}_2$ is given by Eqn. (3-21), where $\mathbf{r}$ is the radius vector from $\boldsymbol{\mu}_1$ to $\boldsymbol{\mu}_2$. Such inter-

$$U = [r^{-3}\boldsymbol{\mu}_1 \cdot \boldsymbol{\mu}_2 - 3r^{-5}(\boldsymbol{\mu}_1 \cdot \mathbf{r})(\boldsymbol{\mu}_2 \cdot \mathbf{r})]\ \Xi_0/4\pi \qquad (3\text{-}21)$$

actions between nuclei are indeed important for the solid state. However, for liquid phase NMR experiments in which molecules are tumbling, the average value of U is zero, both for intermolecular and for intramolecular interactions. This is illustrated for the latter case (using extreme directions for r) in Fig. 3-9. The situation with

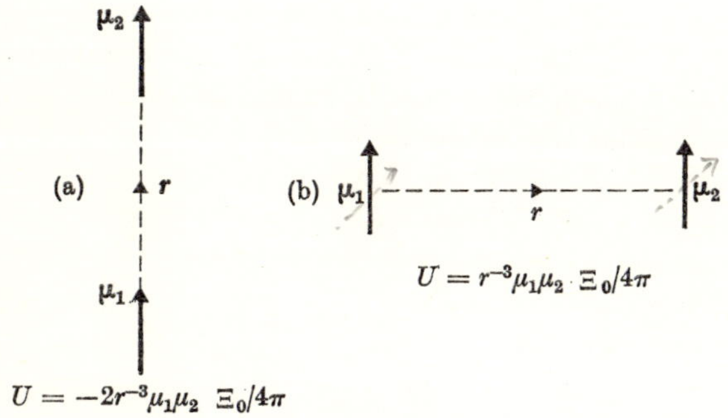

Fig. 3-9 Direct dipolar coupling.

r along the third cartesian axis is equivalent to (b), and so the net energy is $U = \frac{1}{3}[-2r^{-2}\mu_1\mu_2 + r^{-3}\mu_1\mu_2 + r^{-3}\mu_1\mu_2]\ \Xi_0/4\pi = 0$; a more rigorous treatment confirms this result. Nuclear magnetic moments are sufficiently small and internuclear distances sufficiently large for the point-dipole approximation [Eqn. (3-21)] to be valid. Thus, direct nuclear spin-nuclear spin interactions cannot explain the coupling constants observed in high-resolution NMR. The effect must occur indirectly, via the electrons.

3-13 Coupling between electron and nuclear spins

Coupling involving electrons may occur by more than one mechanism First, there is the interaction between the magnetic moments due to nuclear spin and electron orbital motion (the electron-orbital term).

This requires a term in the Hamiltonian of the form of Eqn. (3-22).

$$\hat{\mathscr{H}} = \frac{\Xi_0}{4\pi} g_L \mu_B \gamma_N \hbar \sum_e 2\hat{\mathbf{L}}_e \cdot \hat{\mathbf{I}}_N r_{eN}^{-3} \tag{3-22}$$

where e and N refer to electron and nucleus respectively, $\hat{\mathbf{L}}$ is the orbital angular momentum operator, and g_L is the electron orbital g factor ($= 1$). Secondly there is the dipolar interaction between the electron and nuclear spins (the electron spin-dipole term). The classical energy of interaction is that given by Eqn. (3-21) and therefore the term in the Hamiltonian is

$$\hat{\mathscr{H}} = -\frac{\Xi_0}{4\pi} g_S \mu_B \gamma_N \hbar \sum_e \{r_{eN}^{-3} \hat{\mathbf{I}}_N \cdot \hat{\mathbf{S}}_e - 3 r_{eN}^{-5} (\hat{\mathbf{I}}_N \cdot \mathbf{r}_{eN})(\hat{\mathbf{S}}_e \cdot \mathbf{r}_{eN})\} \tag{3-23}$$

where r_{eN} is the distance between electron e and nucleus N. However, this equation does not completely account for this type of interaction because it assumes point dipoles and therefore is not adequate as $r \to 0$. It does not properly allow for the possibility that the electron may be in the same region of space as the nucleus. Quantum-mechanically this *is* possible so a further energy term is required. This is known as the Fermi contact term. It can be shown to give a contribution (3-24) to the Hamiltonian.

$$\mathscr{H}_{\text{contact}} = \tfrac{2}{3} \Xi_0 g_S \mu_B \gamma_N \hbar \sum_e \delta(r_{eN}) \hat{\mathbf{I}}_N \cdot \hat{\mathbf{S}}_e \tag{3-24}$$

where $\delta(r_{eN})$ is the Dirac delta function (1 if $r_{eN} = 0$ but zero otherwise) and the other symbols have their usual meanings. The use of the delta function gives meaning to the word 'contact' used for this term. In both Eqns. (3-23) and (3-24) the electron spin g-factor, g_S, may be taken as 2.

3-14 The Dirac vector model

Before further discussion of the theory of coupling is given, it may be helpful to take a simple physical picture to show how coupling can occur. However, it must be emphasized that the Dirac vector model mentioned here is a gross oversimplification of the real situation, and many conclusions based on it are found to be incorrect. Equation (3-24) shows that the contact term (which is usually dominant—see the next section) stabilizes the anti-parallel orientation of an electron spin *at* a nuclear spin if γ_N is positive; the electron and nuclear magnetic moments are therefore stable when parallel. Now pairs of electrons in the same orbital have opposed spins according to the Pauli principle, so for HD the stable situation is as in Fig. 3-10(a). This figure assumes that when one of the electrons in a

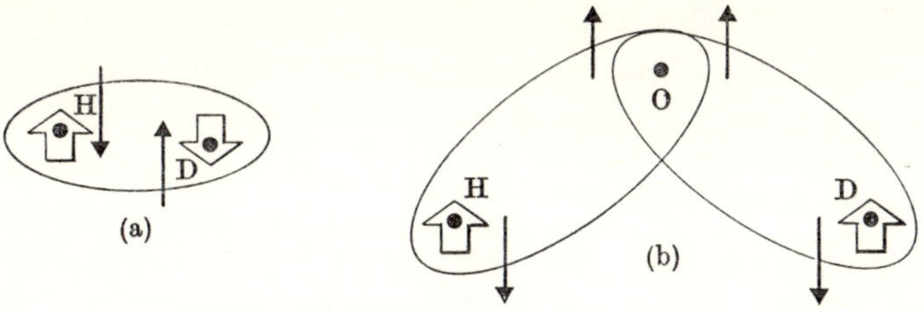

Fig. 3-10 Dirac vector model for proton–proton coupling in (a) HD (b) HDO. The contact term actually involves electrons in the same region of space as the nuclei.

bonding pair is near one of the nuclei there is a tendency for the second electron to be near the second nucleus. As noted in Section 1-12 the NMR energy Eqn. (1-27) is written in such a way that a coupling constant is defined as positive if coupling stabilizes anti-parallel spins. The Dirac vector model of coupling (Fig. 3-10) therefore indicates that the coupling constant should be positive for the directly-bonded nuclei of HD, and, indeed, for any directly-bonded nuclei with positive magnetogyric ratios (in practice not all such coupling constants are positive).

If another atom intervenes between the coupled nuclei, and the bonds are all localized, the interaction between electrons occupying different orbitals in the region of space around this atom must be considered. Hund's rule states that such electrons should have their spins parallel, and Fig. 3-10(b) shows that this implies a negative coupling constant between nuclei separated by two bonds. A logical extension indicates that the sign of coupling constants should alternate as the number of bonds separating the nuclei increases, being positive for an odd number of intervening bonds. This rule is generally obeyed for (H, H) coupling, though there are exceptions. Notice that the model only considers coupling 'through bonds' via σ electrons when bonds are localized. Since there must be a Hund-type interaction at each intervening atom for the transmission of the spin information to occur, the Dirac vector model predicts a rapid attenuation in the magnitude of J as the number of bonds separating the nuclei increases. In fact for saturated compounds (H, H) coupling is only rarely observed when more than four bonds intervene, but the attenuation is less rapid than the Dirac model would predict.

When comparing coupling constants in different molecules (e.g. in a study of substituent effects) it is useful to classify the values according to the number of intervening bonds; this number may be conveniently indicated by means of a superscript prefix to the symbol J. Thus coupling between directly-bonded nuclei is often indicated by ^1J. This classification is occasionally ambiguous for cyclic systems. Coupling through two and three bonds (^2J and ^3J respectively) is often known as *geminal* and *vicinal* coupling.

The (H, H) coupling constant for the hydrogen molecule has been shown to be (see Section 3-16) $^1J_{\mathrm{HH}} = 284$ Hz. It is clear that coupling through more than one bond must involve delocalization of electrons, though not necessarily best described by the model of Fig. 3-10(b). *Geminal* (H, H) coupling constants via an intervening carbon atom are negative in saturated systems, as predicted by the Dirac vector model, but are over an order of magnitude smaller than $^1J_{\mathrm{HH}}$—they range from -9 to -21 Hz. When the intervening carbon atom is sp^2 hybridized, values of $^2J_{\mathrm{HH}}$ are normally positive (the highest value is ca. $+41$ Hz for formaldehyde); this is clearly in contradiction to the vector model. *Vicinal* (H, H) coupling constants (see Section 3-19) are of the same order of magnitude as geminal values, but are invariably positive. In fact frequently $|^2J_{\mathrm{HH}}| < |^3J_{\mathrm{HH}}|$, and most theories show that there are important contributions to $^2J_{\mathrm{HH}}$ that differ in sign and thus tend to cancel. Coupling between protons separated by more then three bonds is only appreciable in special circumstances (see Sections 3-10 and 3-21). Coupling involving nuclei other than protons is somewhat less well understood, partly because there is not so much data and partly because more electrons may be involved.

After dealing with the theory of coupling in more detail, we shall discuss a few selected classes of coupling while completely ignoring others.

3-15 Electronic coupling of nuclear spins

In order to develop the theory of nuclear spin coupling any further it must be recognized that we need to explain an energy term involving the scalar product $\hat{\mathbf{I}}_N \cdot \hat{\mathbf{I}}_{N'}$ for nuclei N and N'. This occurs in an empirical Hamiltonian, as in Eqn. (2-39), which we normally evaluate using nuclear spin wave functions alone. However, it must be possible to equate such a term to others appearing in the theoretical total Hamiltonian which includes terms for all possible physical interactions. Now the physical interactions we are considering are of very small energy compared to the total electronic energy of the system. Thus we may evaluate their effect using perturbation theory. Moreover, we have stated that direct interactions between nuclear spins average to zero for a liquid, so we need to consider the electron–nuclear interactions discussed in Section 3-13. First-order perturbation terms are of the type $\langle\overline{\Psi}|\, \hat{\mathscr{H}}_{eN}\, |\overline{\Psi}\rangle$, where $\hat{\mathscr{H}}_{eN}$ is the perturbation in question and $\overline{\Psi}$ is the molecular wave function; such a term involves only one nucleus and cannot be equated to the coupling term from the empirical Hamiltonian. Therefore, we must turn to second-order perturbation theory, which gives a contribution to the ground-state energy which is of the form (3-25).

$$-\sum_{n\neq 0} \langle\bar{\Psi}_0|\,\hat{\mathscr{H}}_{eN}\,|\bar{\Psi}_n\rangle\langle\bar{\Psi}_n|\,\hat{\mathscr{H}}_{eN}\,|\bar{\Psi}_z\rangle/(E_n - E_0) \qquad (3\text{-}25)$$

In this term $\bar{\Psi}_0$ and $\bar{\Psi}_n$ refer to the ground-state and nth excited-state total wave functions, calculated before including the electron–nuclear perturbation $\hat{\mathscr{H}}_{eN}$; E_0 and E_n are the energies of $\bar{\Psi}_0$ and $\bar{\Psi}_n$. Since this term involves $\hat{\mathscr{H}}_{eN}$ *twice* it will contain contributions involving $\hat{\mathbf{I}}_N \cdot \hat{\mathbf{I}}_{N'}$. The problem of evaluating the coupling constant $J_{NN'}$ therefore reduces to equating corresponding terms in (3-25) and in $\langle\Psi_{\text{nuclear}}|\,\hat{\mathscr{H}}_{\text{NMR}}\,|\Psi_{\text{nuclear}}\rangle$, where $\hat{\mathscr{H}}_{\text{NMR}}$ is given by Eqn. (2-39). Of course J is in principle an anisotropic quantity; we will consider in detail the zz component only, arising from the contact term alone. The treatment discussed above leads to

$$h\langle\Psi_{\text{nuclear}}|\,J_{NN'zz}I_{Nz}I_{N'z}\,|\Psi_{\text{nuclear}}\rangle = -2(\tfrac{2}{3}\Xi_0 g_S\mu_B\hbar)^2\gamma_N\gamma_{N'}$$
$$\sum_{n\neq 0}\langle\bar{\Psi}_0|\sum_e \delta(r_{eN})\hat{I}_{Nz}\hat{S}_{ez}\,|\bar{\Psi}_n\rangle\langle\bar{\Psi}_n|\sum_{e'}\delta(r_{e'N'})\hat{I}_{N'z}\hat{S}_{e'z}|\bar{\Psi}_0\rangle/(E_n - E_0) \qquad (3\text{-}26)$$

The factor 2 appears on the right-hand side because nucleus N may be in either the first or the second bracket of (3-25). Factorization of $\bar{\Psi}$ into nuclear and electronic parts shows that when $\bar{\Psi}_n$ and $\bar{\Psi}_0$ contain the same nuclear part there is a factor on the right-hand side of $\langle\Psi_{\text{nuclear}}|\,I_{Nz}I_{N'z}\,|\Psi_{\text{nuclear}}\rangle$; this factor also occurs on the left-hand-side, giving Eqn. (3-27):

$$J_{NN'zz} = -2h(\Xi_0 g_S\mu_B/3\pi)^2\gamma_N\gamma_{N'}$$
$$\sum_{n\neq 0}\langle\bar{\Psi}_0|\sum_e \delta(r_{eN})\hat{S}_{ez}\,|\bar{\Psi}_n\rangle\langle\bar{\Psi}_n|\sum_{e'}\delta(r_{e'N'})\hat{S}_{e'z}\,|\bar{\Psi}_0\rangle/(E_n - E_0) \qquad (3\text{-}27)$$

where $\bar{\Psi}$ is now purely electronic. Tumbling averages J so that the observed value is $\frac{1}{3}$ the sum of Eqn. (3-27) and corresponding expressions for $J_{NN'xx}$ and $J_{NN'yy}$. The resulting equation is

Ψ̄ = Slater Determinant.

$$J_{NN'} = -\tfrac{2}{3}h(\Xi_0 g_S\mu_B/3\pi)^2\gamma_N\gamma_{N'}$$
$$\sum_{n\neq 0}\langle\bar{\Psi}_0|\sum_e \delta(r_{eN})\hat{\mathbf{S}}_e\,|\bar{\Psi}_n\rangle\cdot\langle\bar{\Psi}_n|\sum_{e'}\delta(r_{e'N'})\hat{\mathbf{S}}_{e'}\,|\bar{\Psi}_0\rangle/(E_n - E_0) \qquad (3\text{-}28)$$

3S states only

It may be noted that the nature of the electron spin operators shows that only triplet excited states need be considered in the summation. The above treatment has been carried out to give J in frequency units (Hz).

Further evaluation of Eqn. (3-28) is possible if details of electronic wave functions are available; a similar treatment of the electron orbital and electron spin-dipole terms may be made. Theoretical calculations of coupling constants have, in fact, been carried out using both valence bond and molecular orbital theory, and qualitative agreement with experimental values is obtained. For HD it has been estimated that the contact, electron spin-dipole and electron orbital terms make contributions to the observed coupling constant (43·5 Hz) approximately in the ratio 1:0·075:<0·013. It is generally agreed that the contact term is dominant for coupling between first-row elements of the periodic table; this is often assumed to be the case for other situations also.

The physical meaning of the use of second-order perturbation theory is clear. The coupling involves two electron–nucleus interactions, one for each of the coupled nuclei. The coupling information is carried by the electrons. In ESR spectroscopy, hyperfine splitting involves a single electron–nucleus interaction. It is therefore not surprising that ESR hyperfine splitting constants are several orders of magnitude larger than NMR coupling constants, nor that the two types of parameter are related quantities. One further note of explanation is necessary regarding the electron spin-dipole term: the first-order contribution of this term to the Hamiltonian averages to zero when molecular tumbling occurs, but the second-order contribution does not. Therefore this mechanism contributes to NMR coupling constants but *not* to ESR splitting constants. Under solution conditions the latter are entirely determined by the contact term.

3-16 The effect of magnetogyric ratios

Equation (3-28) shows that the contact contribution to coupling constants is proportional to the product of the magnetogyric ratios of the nuclei; the same is true of the electron orbital and electron spin-dipole terms. This fact arises because all the interactions are of magnetic origin and involve products of nuclear magnetic moments. Use may be made of this feature in order to evaluate coupling constants between magnetically equivalent nuclei. Such interactions do not normally affect NMR spectra, but isotopic substitution will immediately render nuclei non-equivalent and the analogous coupling constants appear as direct first-order splittings in the spectra. Since isotopic substitution does not affect electronic wave functions, the effect on J is entirely due to the change in magnetogyric ratios. Thus measurement of the spectrum of HD shows that $|^1J_{HD}| = 45{\cdot}3$ Hz; the corresponding value of $|^1J_{HH}|$ is obtained from Eqn. (3-29)

$$J_{H_2}/J_{HD} = \gamma_H/\gamma_D = 6{\cdot}515 \qquad (3\text{-}29)$$

The numerical value of this ratio applies to any H → D substitution. This method is in quite common use—for example, for $^2J_{HH}$ of CH_3 groups the corresponding molecule containing CH_2D is studied (see also Chapter 4 where the proton spectrum of PHD_2 is illustrated). To a first approximation, shielding constants are unaffected by isotopic substitution, so that direct deuterium resonance often gives first-order spectra for fully deuterated molecules even when the corresponding proton resonance shows second-order effects.

It has been suggested that comparison between coupling constants of different elements is only useful when the effect of magnetogyric ratios has been eliminated. Thus it is possible to define *reduced* coupling constants, usually given the symbol K, as follows

$$K_{jk} = \frac{J_{jk}}{h}\left(\frac{\gamma_j}{2\pi}\right)^{-1}\left(\frac{\gamma_k}{2\pi}\right)^{-1} \tag{3-30}$$

This definition is used so that the coupling energy is expressed in terms of nuclear magnetic moments as

$$U_{jk} = K_{jk}\mu_j\mu_k \tag{3-31}$$

On the SI system* the magnetogyric ratios are in rad T^{-1} s^{-1} and K appears in N A^2 m^{-3}. Table 3-3 gives examples for directly-bonded nuclei. There is a general tendency for the magnitude of ^1K to increase as the atomic numbers of the coupled nuclei increase. However, because γ for the proton is high, coupling constants involving

Table 3-3 Coupling constants between directly bonded nuclei

Nuclei	*Molecule*	^1K (10^{20} N A^2 m^{-3})	^1J (Hz)
H, H	H_2	(+)2·3	±276
H, C	CH_4	+4·1	+125
H, O	H_2O	±4·5	±73
H, F	HF	(−)4·6	(−)521
H, Si	SiH_4	+8·5	−202
H, P	PH_3	+3·7	(+)182
H, Sn	SnH_4	+43·0	(−)1,931 (^{1}H, ^{119}Sn)
C, C	CH_3CH_3	+4·6	(+)35
C, F	CF_4	−9·1	−259
C, Si	Me_4Si	+8·7	−52
C, Se	Me_2Se	−10·8	−62
C, Te	Me_2Te	−17·0	+162
C, Hg	Me_2Hg	+128·0	+689
P, P	P_2H_4	−5·5	(−)108
Sn, Sn	Sn_2Me_6	+261·1	+4,264 (^{117}Sn, ^{119}Sn)

* In electromagnetic units (with Ξ_0 considered dimensionless) K is often quoted in cm^{-3}. We have chosen not to use Ξ_0 explicitly in Eqns. (3-30) and (3-31) so that these equations remain as originally defined; however, this means the units of K in SI are *not* m^{-3}.

protons are anomalously large if quoted in Hz. Since magnetogyric ratios may be either positive or negative, the sign of K is not necessarily the same as that of the corresponding J. It has been shown that when the hybridizations are constant, values of $({}^1K_{\mathrm{XH}})^{1/2}$ and $({}^2K_{\mathrm{XCH}})^{1/2}$ are proportional to Z_{X}, the atomic number of X, except for $Z_{\mathrm{X}} < 10$. This occurs because of the dependence of reduced coupling constants on the square of the atomic s-electron charge density at the nucleus (see Section 3-18).

3-17 Coupling between directly bonded nuclei

The contact contribution to coupling depends on the probability of an electron being at the nucleus. However, for atoms only electrons in s orbitals have any electron density at the nucleus. Thus any development of Eqn. (3-28) using molecular orbital theory needs to consider only the contributions of the valence atomic s orbitals for the coupled nuclei. The theory of Pople and Santry shows that if the molecular wave functions Ψ are expressed in terms of molecular orbitals ψ and if the latter are expanded as linear combinations of atomic orbitals ϕ according to Eqn. (3-32)

$$\psi_i = \sum_{\mu} c_{i\mu}\phi_{\mu} \tag{3-32}$$

then the space part (involving δ) and the spin part (involving $\hat{S}$) of Eqn. (3-28) may be evaluated independently and the reduced coupling constant between nuclei j and k may be written

Derive.

$$K_{jk} = (\tfrac{2}{3}\Xi_0\mu_B)^2\langle s_j|\delta(r_j)|s_j\rangle\langle s_k|\delta(r_k)|s_k\rangle\Pi_{s_j,s_k} \tag{3-33}$$

where s_j and s_k are the valence shell s functions for j and k respectively, the integrals $\langle s|\,\delta\,|s\rangle$ represent s-electron density at a nucleus and Π_{s_j,s_k} is known as the mutual polarizability of the orbitals, given by Eqn. (3-34).

$$\Pi_{s_j,s_k} = 4\sum_{i}^{\mathrm{occ}}\sum_{\ell}^{\mathrm{unocc}}(\varepsilon_i - \varepsilon_\ell)^{-1}c_{is_j}c_{is_k}c_{\ell s_j}c_{\ell s_k} \tag{3-34}$$

Only excited states involving the promotion of one electron are considered, and the excitation energy is therefore written as the difference of one-electron energies $(\varepsilon_i - \varepsilon_\ell)$; the coefficients c are those of Eqn. (3-32). In comparing Eqn. (3-32) with Eqn. (3-28) the g-factor has been set at 2 and a further factor of $\frac{3}{2}$ has appeared. The 3 arises from the separate and equal contributions of the three excited triplet states for each excitation energy $(\varepsilon_i - \varepsilon_\ell)$; the factor $\frac{1}{2}$ appears from the combined effects of evaluating the electron spin part of Eqn. (3-28) and of the necessary use of anti-symmetrized molecular wave functions Ψ. Clearly, Eqn. (3-33) describes the mixing of excited states into the ground-state wave function by the magnetic field, as discussed in Section 3-15. Each 'excited state' can be seen to make its own contribution to K; however, the contribution may be positive or negative, depending on the signs of the coefficients in Eqn. (3-34). The value of K itself may be positive or negative—both situations are found experimentally for directly-bonded coupling—in contrast to the predictions of the simple Dirac vector model. The most important 'excitations' are those with small values of $(\varepsilon_i - \varepsilon_\ell)$.

Equation (3-33) is valid for coupling through any number of bonds; we will discuss at this point its application to coupling between directly-bonded nuclei. For tetrahedral molecules or ions such as BH_4^-, CH_4, and NH_4^+, only the two totally symmetrical orbitals (one bonding and one anti-bonding) involve the s wave

Table 3-4 Coupling constants for BH_4^-, CH_4 and NH_4^+

	K_{XH} (10^{20} N A^2 m^{-3}) *Calculated*	*Found*
BH_4^-	+2·2	(+)2·13
CH_4	+4·4	+4·18
NH_4^+	+7·5	(+)5·9

function of the central atom, and therefore only one 'excitation' contributes to K. The product of coefficients for Eqn. (3-34) is positive, and therefore so is K; the magnitude of K is principally determined by the s-electron densities at the nuclei for the valence shell atomic orbitals. These densities increase in the same order as atomic number for the elements considered here (H, B, C, N, O, F). Calculated and observed values of K for the three tetrahedral species are listed in Table 3-4.

For (H, X) coupling in NH_3, H_2O, and HF there are two 'excitations' to consider since three molecular orbitals (two occupied) involve the s function of X. These two excitations give contributions to K of opposite sign (Problem 3-6); when the energy difference between s and p electrons is large, the negative term dominates. Now this energy difference increases in the series C, N, O, F and the calculated sign of K changes in going from K_{CH} to K_{HF}. For the corresponding series of (C, X) coupling constants it has been established experimentally that ${}^1K_{CF}$ is negative whereas ${}^1K_{CH}$ is positive. In fact, all signs so far determined for ${}^1K$ involving fluorine are negative (see Table 3-3).

3-18 Hybridization and related effects

Attention can be focussed to some of the physical concepts behind Eqn. (3-33) if the bond between the coupled nuclei is considered to be of the localized σ type. Then it can be shown that the mutual polarizability reduces to $\prod_{s_j,s_k} = \alpha_j^2\alpha_k^2/\Delta E$ where α_j^2 is the s character of the hybrid orbital used by j in the j, k bond, and ΔE is the (positive) energy of excitation from the bonding to the anti-bonding j, k molecular orbital. The coupling constant may be written:

$$^1J_{jk} \propto \gamma_j\gamma_k\alpha_j^2\alpha_k^2\langle s_j|\ \delta(\mathbf{r}_j)\ |s_j\rangle\ \langle s_k|\ \delta(\mathbf{r}_k)\ |s_k\rangle/\Delta E \qquad (3\text{-}35)$$

It is clear that Eqn. (3-35) cannot account for negative reduced coupling constants, but it is useful for predicting certain variations in J. In particular Eqn. (3-35) shows that coupling constants should have a pronounced dependence on the s character of the σ-bond between the nuclei, i.e. on hybridization. This is well illustrated by $^1J_{CH}$ (see Table 3-5), though the accuracy of the proportionality may be fortuitous.

Table 3-5 Dependence of $^1J_{CH}$ on carbon hybridization

Molecule	*Hybridization*	α_C^2	J (Hz)
Ethane	sp^3	$\frac{1}{4}$	124·9
Ethylene	sp^2	$\frac{1}{3}$	156·4
Acetylene	sp	$\frac{1}{2}$	248·7

At a more empirical level it has been noted that for methane derivatives, CHXYZ, there are additive substituent contributions, ξ, such that the (C, H) coupling constants may be expressed as:

$$^1J_{CH} = \xi_X + \xi_Y + \xi_Z \qquad (3\text{-}36)$$

There are deviations for multiple substitution by highly electronegative groups and extra terms must be introduced into Eqn. (3-36) to account for pairwise interactions between substituents. The parameters ξ correlate with the electronegativity of the substituent, but this affects the s-electron density at the carbon atom in at least two ways. First, the hybridization of the carbon atom used in the various bonds (i.e. α^2) is affected: electronegative substituents prefer to make use of carbon p character, since the electron density is at a maximum further from the carbon nucleus than for the carbon s orbital. This results in an increase in the s character to the remaining substituents (a lone pair will occupy an orbital of high s character) and the coupling constant will therefore increase. Secondly, the bond to an electronegative substituent is polarized. The electron density around carbon is reduced, but this results in an increased effective nuclear charge for the carbon nucleus so that the radial electron distribution is modified. There is a contraction of the electron cloud of carbon; $\langle s_C | \delta(r\ ) | s_C \rangle$ would increase if α^2 remained constant and therefore coupling constants increase (see Eqn. 3-35). Both effects act in the same direction and it is not clear which is dominant. It is certainly unwise to derive s characters from values of J_{CH}. However, in the case of the platinum complex *cis*-[PtClMe-$(Et_3P)_2$], the two chemically shifted phosphorus resonances have very different coupling constants to platinum ($^1J_{PtP} = 4{,}179$ Hz and 1,719 Hz). In this situation $\langle 6s_{Pt} | \delta(r_{Pt}) | 6s_{Pt} \rangle$ is common to the two cases and the observed differences must be due to different values of α^2_{Pt} (provided the contact term is dominant and π-type electrons can be ignored).

3-19 *Vicinal* (H, H) coupling

Valence bond theory has been of great success in qualitatively describing trends in $^3J_{HH}$. The situation is clearly rather complex

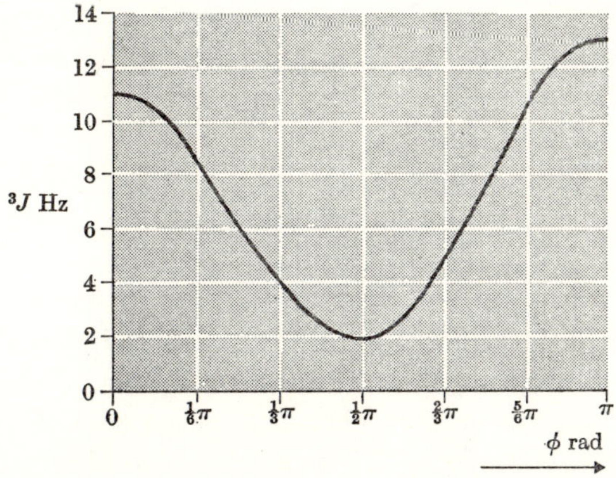

Fig. 3-11 Variation of *vicinal* (H, H) coupling constants in hydrocarbons with dihedral angle.

because of the considerable number of electrons and geometrical parameters involved, so we will give no details of the theory. The most important feature is that, when other factors are constant, the *vicinal* coupling constant through carbon is predicted to depend on the dihedral angle, ϕ, between the C—H bonds as in Eqn. (3-37)

$$^3J = A + B\cos\phi + C\cos 2\phi \qquad (3\text{-}37)$$

where A, B, and C are constants with approximate values 4·0 Hz, −0·5 Hz, and 4·5 Hz respectively.* However, use of these values

* This equation is sometimes written in the form $^3J = D\cos^2\phi + E$, where D is somewhat larger for $\pi/2 \leqslant \phi \leqslant \pi$ than for $0 \leqslant \phi \leqslant \pi/2$.

usually underestimates ^3J, and a better empirical set for hydrocarbons is found to be $A = 7$ Hz, $B = -1$ Hz, $C = 5$ Hz. This variation with ϕ is shown in Fig. 3-11. Equation (3-37) is known as the Karplus equation, and has been used to predict dihedral angles in compounds of unknown conformation. However, there are considerable dangers in this procedure since the Karplus model is only valid in the absence of electronegative substituents and of departure from tetrahedral angles at carbon. Thus qualitative use of Eqn. (3-37) is useful, particularly when values of A, B, and C are determined from model compounds closely related to the unknown, whereas quantitative use is usually to be avoided. Experiment shows that *vicinal* (H, H) coupling constants increase in the order *gauche* ($\phi = \pi/3$) < *cis* ($\phi = 0$) < *trans* ($\phi = \pi$) in the absence of other influences than ϕ. Some typical values are given for schemes [**9**], [**10**,] and [**11**]. Only values averaged over internal rotation about the C—C bond are normally obtained for acyclic ethane derivatives;

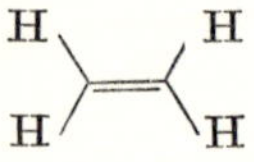

$J_{cis} = 11{\cdot}6$ Hz ($\phi = 0$)
$J_{trans} = 19{\cdot}1$ Hz ($\phi = \pi$)

[**9**]

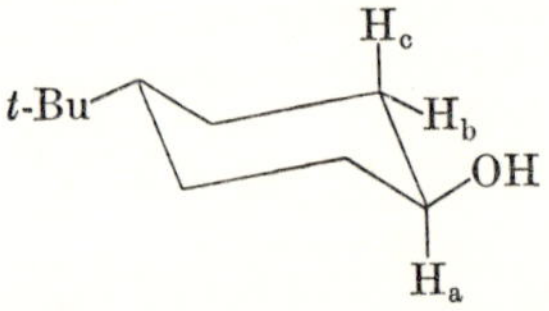

$J_{ab} = 4{\cdot}31$ Hz ($\phi \sim \pi/3$)
$J_{ac} = 11{\cdot}07$ Hz ($\phi \sim \pi$)

[**10**]

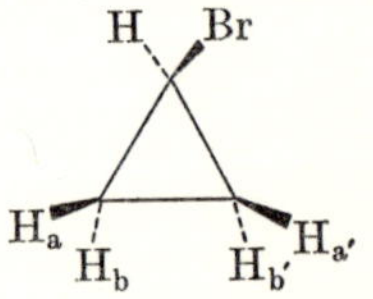

$J_{aa'} = 11{\cdot}0$ Hz ($\phi \sim 0$)
$J_{ab'} = 7{\cdot}5$ Hz ($\phi \sim 2\pi/3$)

[**11**]

for compounds of the type CH_3CHXY the observed value is $\frac{2}{3}J_g + \frac{1}{3}J_t$ where J_g and J_t are the *gauche* and *trans* coupling constants respectively. The situation is more clear-cut in the case of cyclohexane derivatives in the chair form. There are five possible mutual orientations of the protons, each of which may be equatorial (e) or axial (a)—see Table 3-6. The first three rows of this table are for 'fixed'

Table 3-6 Orientations and *vicinal* coupling constants for cyclohexane derivatives

Position of protons	ϕ	*Expected* ^3J (Hz)[a]
aa	π	13
ae	$\pi/3$	4
ee	$\pi/3$	4
aa ⇌ *ee*	$\pi \rightleftharpoons \pi/3$	8·5
ae ⇌ *ea*	$\pi/3 \rightleftharpoons \pi/3$	4

(a) Using Eqn. (3-37) with $A = 7$ Hz, $B = -1$ Hz, $C = 5$ Hz.

chair conformations, which occur when the two possible chair forms are non-equivalent (or at low temperatures in cases where they are equivalent); the last two rows apply when rapid ring inversion occurs, as is true at room temperature when the inversion isomers are equivalent. In practice, $^3J_{ee}$ is generally found to be less than $^3J_{ae}$, particularly when electronegative substituents are present. Chemical applications of this type of information (typical of the use of NMR in organic chemistry) may be illustrated with four of the five possible isomers of 2,3,5,6-tetramethylpiperazine. The value of ^3J for the ring protons of the dihydrochlorides are given in Fig. 3-14 underneath the conformations deduced from this data. Actually, the assignment of the γ and δ isomers cannot be made from this data alone, as can be seen.

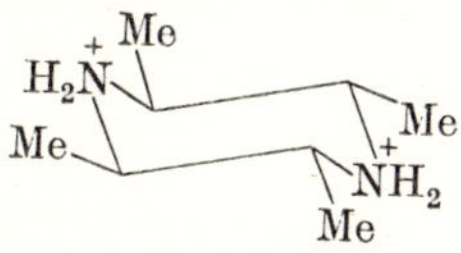

J_{aa}
11·1 Hz
α isomer
[**12**]

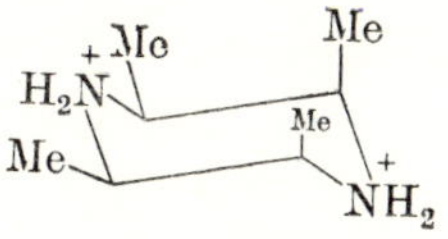

$J_{ae} \rightleftharpoons J_{ea}$
4·3 Hz
γ isomer
[**13**]

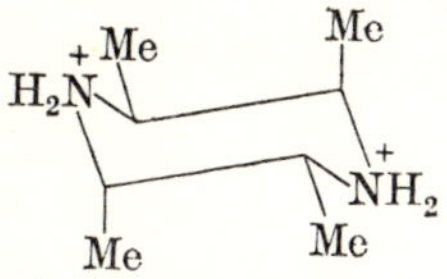

$J_{ae} \rightleftharpoons J_{ea}$
4·1 Hz
δ isomer
[**14**]

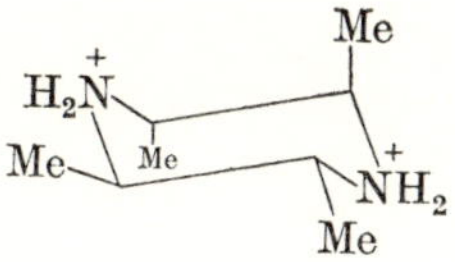

$J_{aa} \rightleftharpoons J_{ee}$
6·6 Hz
ε isomer
[**15**]

Valence bond calculations also make the following predictions, which agree with experimental observation—though the effects are not independent:

(a) Electronegative substituents decrease ^3J.
(b) Increase of HCC bond angles decreases ^3J.
(c) Increase of C—C bond length decreases ^3J.

Effect (a) is well illustrated by ethyl compounds and may be expressed by Eqn. (3-38), where $^3J^0$ is the value for ethane itself (8·0 Hz)

$$^3J(\mathrm{CH_3CH_2X}) = {}^3J^0[1 - 0{\cdot}09(E_\mathrm{X} - E_\mathrm{H})] \qquad (3\text{-}38)$$

while E_X and E_H are the electronegativities of the substituent X and hydrogen respectively. However, the magnitude of the electronegativity effect depends on the dihedral angle, and empirically it is found that the effect is greatest (i.e. $^3J_{\mathrm{HH}}$ least) when X is *trans* to one of the coupled protons. Some examples will make this clear.

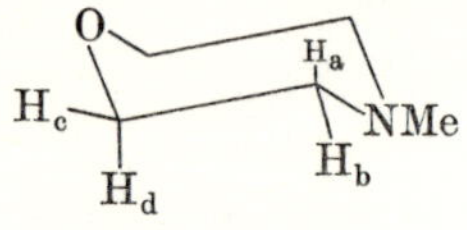

$J_{bd} = 2{\cdot}16$ Hz
$J_{ac} = 2{\cdot}96$ Hz
$J_{bc} < 1{\cdot}0$ Hz

[16]

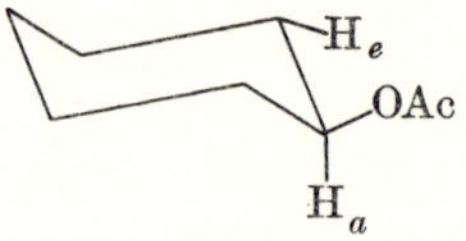

$J_{ae} \sim 5{\cdot}5$ Hz

[17a]

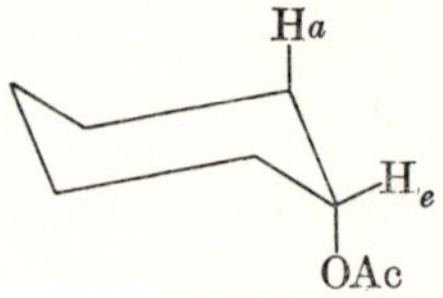

$J_{ae} \sim 3{\cdot}0$ Hz

[17b]

The sign of $^3J_{\mathrm{HH}}$ is invariably found to be positive, in agreement with the simple Dirac vector model. Indeed, relative signs of other coupling constants are often based on an assumption that $^3J_{\mathrm{HH}}$ is

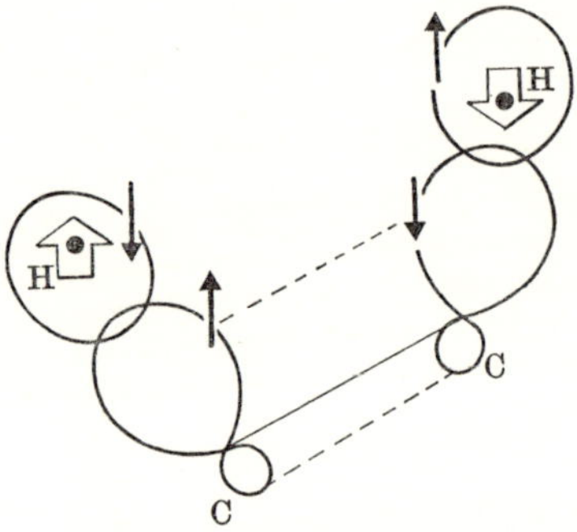

Fig. 3-12 Spin correlation for *vicinal* (H, H) coupling.

positive. *Vicinal* coupling constants are also normally positive for other nuclei coupled to hydrogen, although negative values of $^3J_{\mathrm{FF}}$ are well-known. However, if the Dirac vector model were entirely valid, no dependence of the coupling constant on dihedral angle would be expected. In fact, the most important exchange integral determining the electron spin correlation between electrons in different bonds is that between *vicinal* orbitals (Fig. 3-12). This causes the electrons to be stable when anti-parallel and thus leads to a positive *vicinal* coupling constant (as does the Dirac vector model). Figure 3-12 makes it clear that the electron correlation depends greatly on dihedral angle (since overlap of the *vicinal* orbitals does), and indicates that it will be optimum when the bonds are *cis* or *trans*. In fact the overlap (and hence ^3J) is slightly greater for *trans* bonds than for *cis* bonds.

3-20 Long-range (H, H) coupling

Coupling over more than three chemical bonds is normally termed 'long-range'. For protons in saturated compounds such coupling is normally small (often negligible). Moreover, it is often highly stereospecific; the optimum path for such coupling appears to be a planar zig-zag atomic chain (schemes [**18**] and [**19**]). Some examples are given in schemes [**20–25**]. Note that $|^4J|$ increases rapidly as the number of 'coupling paths' increases. One of the commonest favourable situations is that of coupling over four bonds between equatorial

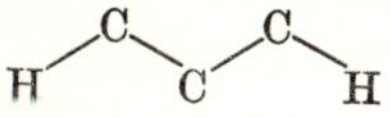

^4J

[**18**]

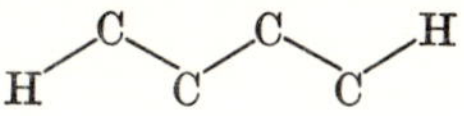

^5J

[**19**]

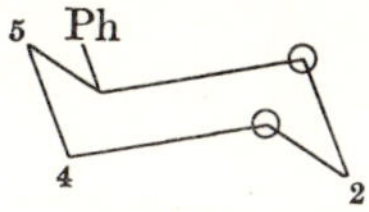

$|^4J_{2e,4e}| = 1{\cdot}0$ Hz $|^4J_{2e,4a}| = 0{\cdot}45$ Hz
$|^5J_{2e,5e}| = 0{\cdot}9$ Hz $|^5J_{2e,5a}| < 0{\cdot}3$ Hz

[**20**]

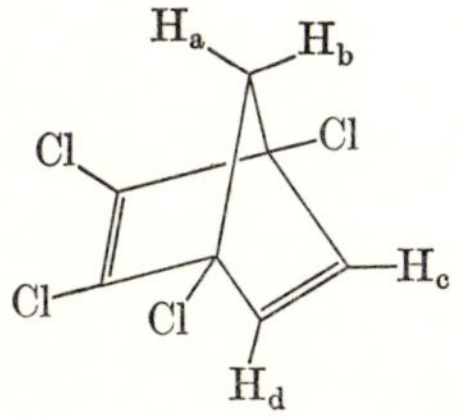

$^4J_{ac} = {^4J_{ad}} = +0{\cdot}64$ Hz
$^4J_{bc} = {^4J_{bd}} = -0{\cdot}15$ Hz

[**21**]

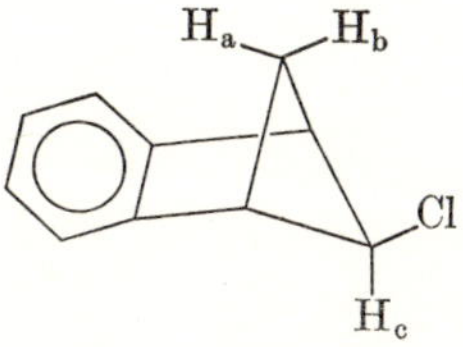

$J_{ac} = +7{\cdot}4$ Hz
$J_{bc} = -0{\cdot}2$ Hz

[**22**]

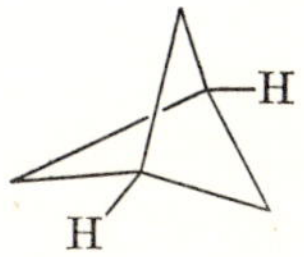

$|^4J| = 18$ Hz

[**23**]

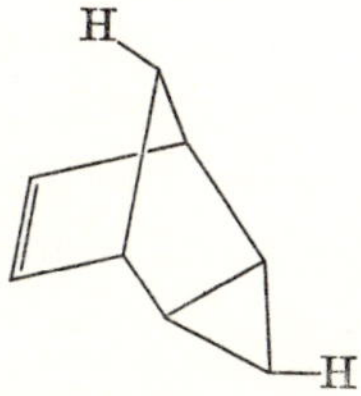

$|^5J| = 2{\cdot}3$ Hz

[**24**]

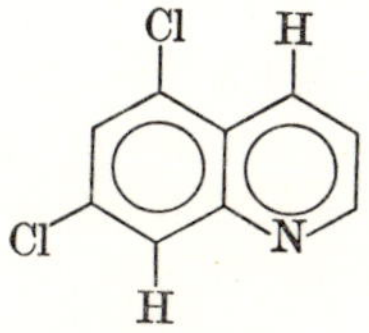

$|^5J| = 0{\cdot}8$ Hz

[**25**]

protons in six-membered saturated rings. Another favourable situation for ^4J occurs when the intervening atom is sp^2 hybridized, as in acetone and dimethylsulphoxide ($|^4J| = 0{\cdot}55$ Hz and $0{\cdot}45$ Hz, respectively, for the neat liquids). Normally, however, the coupling for acyclic compounds is averaged over several conformations, few of which have the optimum geometry, so the observed value is usually very small or undetectable. It may be noted that both positive and negative values of $^4J_{HH}$ have been reported, but the value for the planar zig-zag relationship (when π effects are absent) is probably always positive.

Both the stereospecificity and the frequent occurrence of positive signs indicate that the Dirac vector model cannot adequately account for the observed values of ${}^4J_{HH}$. The stereospecificity indicates that *vicinal* exchange interactions (i.e. direct spin correlation between electrons in *vicinal* orbitals) are again of importance. It has been suggested that at least two types of coupling path must be considered. These may be described in terms of spin correlation between electrons in different bonds (Fig. 3-13). Path (a) involves two *vicinal* orbital interactions (see Fig. 3-12) and gives a negative contribution to ${}^4J$, whereas path (b) involves one *vicinal* orbital interaction and one

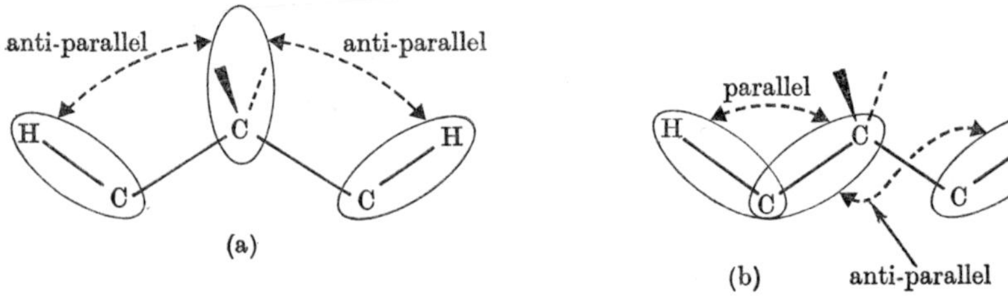

Fig. 3-13 Coupling paths for ${}^4J_{HH}$.

geminal interaction [spins parallel as for Hund coupling—see Fig. 3-10(b)] leading to a positive contribution to ${}^4J$. The total coupling constant may therefore be positive or negative; since the *vicinal* orbital interactions are maximum when the orbitals are *trans*, path (b) dominates for the zig-zag conformation, giving a positive sign (and, in fact, maximum magnitude) in this case. The Dirac vector model provides a further coupling path which gives a negative contribution to ${}^4J_{HH}$, independent of conformation.

3-21 Coupling through π electrons

The π electrons appear to be involved in the nuclear spin–spin coupling for unsaturated compounds; in fact the observed coupling constant may be written as the sum of two parts, J^σ and J^π. Of course, the π-electron distributions have nodes at the nuclei and so cannot contribute to the contact term directly. Consequently it is necessary to introduce spin correlation between σ and π electrons. This same interaction accounts for hyperfine coupling between protons and unpaired electrons in the ESR of related free radicals. It is possible to obtain a relation between J and the hyperfine splitting constants. Three types of σ–π exchange repulsion may be considered:

(a) acetylenic—the σ electron on H is stabilized anti-parallel to the π electron on C for a H—C≡ group.

(b) ethylenic—again the anti-parallel orientation is stabilized.

(c) allylic—in this case the parallel orientation is stable.

Figure 3-14 shows the stable situation for propene, involving both ethylenic and allylic interactions. It may be seen that a negative value of ^4J is predicted. In fact the allylic σ–π exchange repulsion depends on the dihedral angle about the C—C single bond, being approximately zero when all the nuclei involved are coplanar, and maximum when the allylic C—H bond is perpendicular to the

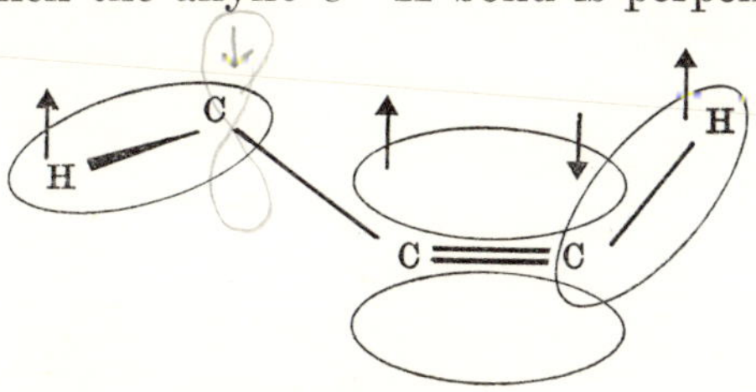

Fig. 3-14 Spin correlation for allylic $^4J^{\pi}_{HH}$.

C═C plane. For allyl compounds the observed coupling constant is a weighted average over all conformations. Further conclusions of the theory are:

(i) ^3J (ethylenic) should be positive, ^4J (allylic) should be negative, and ^5J (known as homoallylic for the group HC—C═C—CH) should be positive.

(ii) Orientation about a double bond should not affect J; e.g. the *cisoid* ^4J illustrated in Fig. 3-14 should equal the corresponding *transoid* ^4J.

(iii) Since π-electron systems are often delocalized, spin correlation may occur for electrons at atoms separated by many chemical bonds, i.e. π-coupling should be long-ranged. The magnitude of coupling constants in conjugated systems may be evaluated using linear combinations of atomic orbitals in a similar fashion to that developed in Section 3-17.

These conclusions agree in general with experimental observation, though it is often difficult to estimate J^{π} because of contributions of unknown magnitude from J^{σ}. Values of $^3J_{HH}$ are dominated by the σ contribution; both σ and π effects may be important for ^4J and ^5J. Table 3-7 lists some values of (H, H) coupling constants in unsaturated systems. The signs, where known, are in agreement with prediction (i) above. Orientation about the double bond seems to affect ^4J and ^5J in *cis*- and *trans*-2-butene to only a small extent. The

Table 3-7 (H, H) coupling constants in some unsaturated compounds

Compound	^4J	^5J	^6J	^7J
cis-2-Butene	−1·79	+1·18	–	–
trans-2-Butene	−1·73	+1·60	–	–
Allene	−7·0	–	–	–
Butatriene	–	±8·95 (*cis*)	–	–
		±8·95 (*trans*)	–	–
1,3-Butadiene	−0·83 (*cis*)	+0·69 (*cis,cis*)	–	–
	−0·86 (*trans*)	+0·60 (*cis,trans*)		
		+1·30 (*trans,trans*)		
cis-1,3-Pentadiene		+0·21 (*cis* H, CH_3)	−0·59 (*cis,cis* H, CH_3)	–
			−0·61 (*cis,trans* H, CH_3)	
cis,cis-2,4-Hexadiene		+0·32 (*cis* H, CH_3)	−0·70 (*cis,trans* H, CH_3)	+0·47 (*cis,cis* CH_3, CH_3)

values of ^5J for butatriene are large due to the existence of several paths for π-coupling; $^5J_{cis}$ and $^5J_{trans}$ are equal within experimental error, thus indicating that the π contribution is dominant. The case of 1,3-butadiene is of considerable interest; of the three different coupling constants over five bonds, the values for *cis-cis* and *cis-trans* orientations are approximately equal (0·69 Hz and 0·60 Hz respectively) and are therefore suggested to be dominated by J^π, whereas the *trans-trans* value is considerably larger (1·30 Hz) indicating a contribution from J^σ. This is not unexpected since the coupling path is of the planar zig-zag type known to be favourable for σ coupling (see preceding section).

The ESR hyperfine splitting constant of the ethylenic type is in fact somewhat smaller in magnitude (−65 MHz) than the allylic constant averaged over all the positions of a methyl group, (+75 MHz). Consequently, replacement of an ethylenic proton by a methyl group should affect π coupling to any other proton by (a) a slight increase in magnitude, and (b) a change of sign. This prediction has been used to test whether coupling is of the π-type or not. Table 3-7 plus related data shows that for the butadiene series such a test confirms that $^5J_{cis,cis}$ and $^5J_{cis,trans}$ are π-dominated whereas $^5J_{trans,trans}$ has roughly equal contributions from σ and π mechanisms.

The long-range nature of J^π is well illustrated by the compound $CH_3C{\equiv}CC{\equiv}CC{\equiv}CCH_2OH$, where an observable coupling (0·4 Hz) is obtained through no fewer than nine chemical bonds.

3-22 'Through-space' coupling

In certain instances, anomalously large coupling constants are observed for nuclei many bonds apart but physically close in space. This occurs particularly for (F, F) coupling but has been found in other cases also. Some representative examples are shown below:

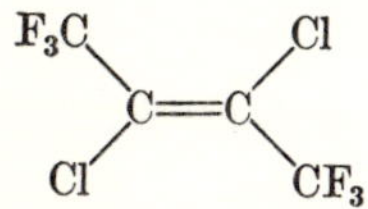

$^5J_{FF} = 1{\cdot}44$ Hz

[26]

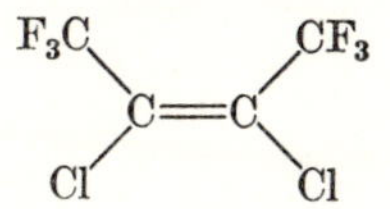

$^5J_{FF} = 13{\cdot}39$ Hz

[27]

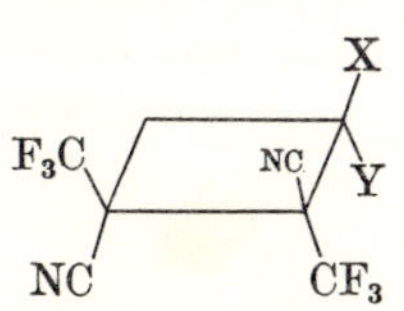

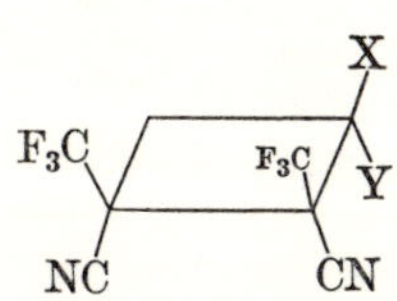

X = H, Y = OEt and vice versa

$J_{FF} <$ ca. 1 Hz $\quad$ $J_{FF} = 10$ to 12 Hz

[28] $\quad$ **[29]**

It should be noted that there is some dispute about the existence and origin of this effect. However, it is not necessary to postulate any novel coupling mechanism; all that is implied is that there is direct electron spin correlation between electrons of fluorine nuclei (or of bonds to such nuclei) close in space. The name 'through-space' is therefore rather misleading. 'Through-space' coupling may be the dominant effect for (F, F) interaction in saturated compounds when the nuclei are separated by five or more chemical bonds.

It has been suggested, from both theoretical and experimental standpoints, that 'through-space' effects may also be noticeable for (H, F) and (H, H) coupling constants. The contribution to J_{HH} from this effect is usually positive. Such an effect provides a fourth coupling path for $^4J_{HH}$ (in addition to those discussed in Section 3-21) which clearly is highly dependent on conformation.

Problems

3-1 By transformation to polar coordinates, show that Eqn. (3-7) reduces to (3-8) for a spherical charge distribution.

3-2 From the data of Table 3-2 calculate the angle which the nodal surface of shielding (due to magnetic anisotropy) makes with the bond direction in the nodal plane of the π electrons for the carbonyl group.

3-3 Show that Eqn. (3-14) reduces to (3-15) in all cases, and becomes (3-16) if $\chi'_{\perp} = \chi''_{\perp}$.

3-4 Show that if $[D]_0 \gg [A]_0$, Eqn. (3-20) applies to NMR measurements for a charge-transfer equilibrium.

3-5 The hydrogen molecule has molecular orbitals $\psi_g = (s_A + s_B)/\sqrt{2}$ and $\psi_u = (s_A - s_B)/\sqrt{2}$, where s_A and s_B denote 1*s* atomic orbitals on the two hydrogen atoms A and B. The ground and the first excited electronic states, Ψ_0 and Ψ_1 respectively are

$$\Psi_0 = \psi_g(\mathbf{r}_1)\psi_g(\mathbf{r}_2)[\alpha_1\beta_2 - \beta_1\alpha_2]/\sqrt{2}$$

$$\Psi_1\ (m_s = 0 \text{ component}) = \tfrac{1}{2}[\psi_g(\mathbf{r}_1)\psi_u(\mathbf{r}_2) - \psi_u(\mathbf{r}_1)\psi_g(\mathbf{r}_2)][\alpha_1\beta_2 + \beta_1\alpha_2].$$

Evaluate the contact contribution to the (H, H) coupling constant using Eqn. (3-28) and the following information:

$$E_1 - E_0 = 1{\cdot}6 \times 10^{-18}\ \text{J}$$

$$\langle s_j|\ \delta(r_j)\ |s_j\rangle = 2{\cdot}15 \times 10^{30}\ \text{electrons m}^{-3}$$

(Evaluate the S_z component and multiply by 3 to take account of S_x and S_y contributions.)

3-6 Consider the three molecular orbitals of HF obtained from the hydrogen 1*s* and the fluorine 2*s* and $2p_z$ atomic orbitals (z being the bond direction). Take in turn the hypothetical cases where the energies of the orbitals are (a) $H_{1s} \sim F_{2p_z} \gg F_{2s}$, and (b) $H_{1s} \gg F_{2p_z} \gg F_{2s}$ and estimate which 'excitation' is most important for (H, F) coupling. Give the predicted sign of J_{HF} in these two cases.

What is the significance of the following results?

3-7 In dilute solution in CCl_4 the hydroxyl protons of phenol and methyl salicylate give NMR signals at $\delta = 4{\cdot}24$ and $\delta = 10{\cdot}58$ respectively.

3-8 The resonances of the central methylene protons of [**30**] and [**31**] occur at very different frequencies. Which has the higher δ value?

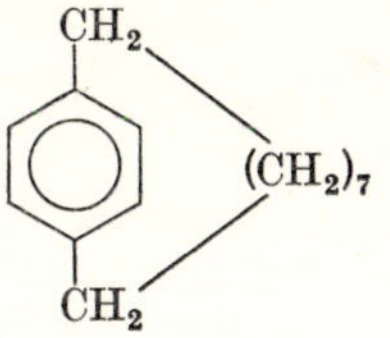

$Ph(CH_2)_9Ph$

[**30**] [**31**]

3-9 The solvent shifts of equatorial protons in the 2- and 3-positions of cyclohexanone are $\delta_{C_6H_6} - \delta_{CDCl_3} = 0{\cdot}03$ and $-0{\cdot}35$ respectively.

3-10 The ^{19}F chemical shifts (from the resonance of $CFCl_3$, with positive values to low frequency) of CF_3Cl and CF_3I are 28·1 and 4·8 p.p.m. respectively.

3-11 $Me_3SnSnMe_3$ has $^2J(^{117}Sn, ^1H) = 47{\cdot}0$ Hz, $^2J(^{119}Sn, ^1H) = 49{\cdot}6$ Hz.

3-12 *Vicinal* coupling constants for ring protons of [**32**] and [**33**] are as below (in Hz):

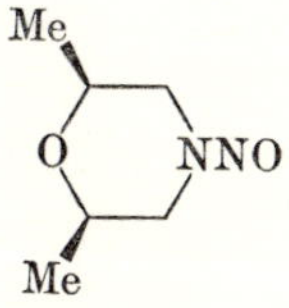

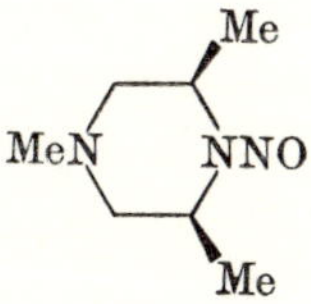

10·7 and 2·8 4·6 and 1·2

[**32**] [**33**]

3-13 2,4-disubstituted benzaldehydes [**34**] have long-range (H, H) coupling constants (in Hz) as follows:

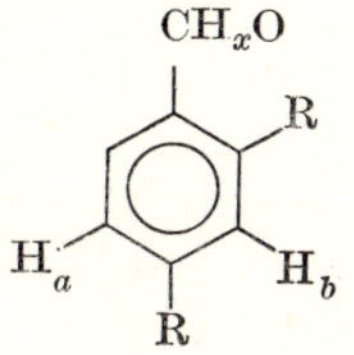

$R = OH$ $J_{ax} < 0{\cdot}1$ $J_{bx} = 0{\cdot}51$

$R = NO_2$ $J_{ax} = 0{\cdot}57$ $J_{bx} = 0{\cdot}21$

[**34**]

4 Relaxation and double resonance

4-1 Introduction

In the previous chapters we have discussed how observed nuclear resonance spectra can be related to nuclear spin energy levels which in turn depend on the molecular wave function. As in other branches of molecular spectroscopy, the frequencies at which electromagnetic radiation is absorbed depend on the separation of the energy levels. Apart from the difference in the type of energy level, the main contrast with microwave, infrared, or ultraviolet spectroscopy lies in the necessity for an external magnetic field. In addition, we have seen that the transitions are induced by the magnetic part of the electromagnetic field rather than the electric part.

In this chapter we shall deal with effects which, although they are not confined to nuclear resonance, are easily observable in this type of spectroscopy. Because differences in energy between nuclear spin states are small at normal temperatures, these states are nearly equally populated; for example, in chloroform there is only 1 proton in 100,000 more in the lower state than in the upper in a field of 2·35 T. The oscillating field induces transitions upwards and downwards with equal probability; the net absorption of energy is due to the small difference in population of upper and lower levels.

As the quanta are small, it is easy to apply a high intensity of coherent electromagnetic radiation to the system. This leads to multiple quantum and other non-linear effects normally associated with lasers. These are discussed in Sections 4-9 to 4-13.

The act of energy absorption reduces the population difference, and, if the power in the radio frequency field is too high, the population difference decreases, weakening the spectrum. This phenomenon, which is sometimes a limiting factor in an attempt to observe weak signals, is known as saturation and is discussed in Section 4-5.

As line widths and line separations are so much less than in other types of spectroscopy it is also possible to detect a variety of dynamic processes which occur in molecules at these frequencies. Examples of chemical exchange are discussed in Section 4-8.

PART A LINE SHAPES AND RELAXATION

4-2 The Bloch equations

[Cf Newton's Principia : The Precession of the Equinoxes].

The pioneers of nuclear resonance were mainly concerned with dynamic processes and line shapes. Indeed, the first observations of chemical shifts and coupling constants in liquids were unexpected. Bloch's approach was macroscopic and classical and is complementary to the use of a spin Hamiltonian. The Bloch equations which we shall derive in this section can be used to discuss the line shape of a simple spectrum, but they are not applicable to a complex spectrum.

In order to derive the equations, we discuss the behaviour of a macroscopic sample containing many identical molecules, each with one magnetic nucleus. The total magnetic moment or magnetization* $\mathbf{M}$ of the sample is the resultant of the nuclear moments $\boldsymbol{\mu}$ which in turn are related to the expectation values of the nuclear spin angular momenta $\hbar\langle\hat{\mathbf{I}}\rangle$,

$$\mathbf{M} = \sum_j \boldsymbol{\mu}_j = \gamma\hbar \sum_j \langle\hat{\mathbf{I}}_j\rangle \tag{4-1}$$

or in terms of the total spin angular momentum $\mathbf{P}(= \sum \hbar\langle\hat{\mathbf{I}}_j\rangle)$

$$\mathbf{M} = \gamma\mathbf{P} \tag{4-2}$$

As the sample is macroscopic, the effect of an applied magnetic field $\mathbf{B}$ can be predicted from classical mechanics. The interaction of the field and moment gives a torque on the system, changing the angular momentum $\mathbf{P}$.

$$\frac{d}{dt}\mathbf{P} = -\mathbf{B} \wedge \mathbf{M} \tag{4-3}$$

but as $\mathbf{P}$ and $\mathbf{M}$ are related by Eqn. (4-2),

$$\frac{d}{dt}\mathbf{M} = -\gamma\mathbf{B} \wedge \mathbf{M} \tag{4-4}$$

The classical motion of the nuclear moments $\boldsymbol{\mu}$ discussed in Section 1-6 is described by an equation of motion similar to (4-4). The steady state solution is a precession of the system about the field direction with angular velocity $\boldsymbol{\omega} = -\gamma\mathbf{B}$. This is known as Larmor precession and the corresponding frequency $\gamma B/2\pi$ is the Larmor frequency.

Any system with a magnetic moment will show Larmor precession. The magnetic moment of the sample in nuclear resonance may, in addition, change by internal realignment of the individual nuclear spins. Indeed, when the sample is placed in a magnetic

* The total magnetization should be distinguished from the magnetization per unit volume or per mole.

field its moment changes from zero to $\mathbf{M}_0$, the equilibrium value. The magnitude of $\mathbf{M}_0$ depends on the temperature T and the field. For spin $\frac{1}{2}$ nuclei this can be calculated by using the Boltzmann distribution to find the difference in population $(n_\alpha - n_\beta)$ of the two levels. If the z direction is chosen along $\mathbf{B}_0$, $\mathbf{M}_0$ is in the same direction and is given by

$$\mathbf{M}_0 = \tfrac{1}{4}N(\gamma\hbar)^2\mathbf{B}_0/kT \tag{4-5}$$

where N is the total number of spins and the high-temperature approximation $kT \gg |\gamma B_0|$ has been used (see Problem 4-1).

The equation of motion for the magnetic moment of the sample must include the approach to thermal equilibrium as well as Larmor precession. Bloch assumed that the components of $\mathbf{M}$ decay to $\mathbf{M}_0$ exponentially, but he allowed the components of $\mathbf{M}$ parallel and perpendicular to $\mathbf{M}_0$ to decay with different time constants T_1 and T_2, so that with the z axis chosen along $\mathbf{B}_0$

$$\frac{d}{dt}M_z = -(M_z - M_0)/T_1 \tag{4-6}$$

$$\frac{d}{dt}M_x = -M_x/T_2 \qquad \frac{d}{dt}M_y = -M_y/T_2 \tag{4-7}$$

The approach to thermal equilibrium is known as relaxation and T_1 and T_2 are relaxation times. The decay of the longitudinal component M_z may differ from the decay of the transverse components M_x and M_y because the energy of the spin system depends on M_z. Any change in M_z is accompanied by an energy flow between the nuclear spin system and the other degrees of freedom of the system known as the 'lattice.' The relaxation time T_1 which describes this flow is sometimes known as the *spin–lattice relaxation time*. An alternative name is the *longitudinal relaxation time*, emphasizing the relationship of M_z and the applied field. T_2 is known as the *transverse relaxation time* or the *spin–spin relaxation time*. The latter name arises because direct interactions between the spins of different nuclei can cause relaxation of M_x and M_y without energy transfer to the lattice. In high-resolution nuclear resonance of fluids, however, such processes have a negligible effect, and T_1 and T_2 in fact both arise from energy exchange between the spin system and the lattice. Section 4-3 contains a further discussion of these processes.

In the nuclear resonance experiment there is a small oscillating field, $2B_1 \cos \omega t$, in the x direction. This can be resolved into two components rotating in opposite directions with angular velocities $\pm\omega$. These can be considered independently provided B_1 is small, and the component rotating in the opposite direction to the Larmor precession (the one with angular velocity $+\omega$) can be neglected. The rotating field induces a rotating magnetization in the x, y plane which is monitored by the spectrometer. The field **B** acting on the sample is

$$\mathbf{B} = B_1 \cos \omega t\, \mathbf{i} - B_1 \sin \omega t\, \mathbf{j} + B_0\, \mathbf{k} \tag{4-8}$$

where **i**, **j**, and **k** are unit vectors in the x, y, and z directions.

Combining Eqns. (4-4) for the Larmor precession and (4-6) and (4-7) for the relaxation with this value of **B**, we obtain the Bloch equations

$$\begin{aligned}\frac{dM_z}{dt} &= -\gamma[B_1 \cos \omega t M_y + B_1 \sin \omega t M_x] - (M_z - M_0)/T_1 \\ \frac{dM_y}{dt} &= -\gamma[B_0 M_x - B_1 \cos \omega t M_z] - M_y/T_2 \\ \frac{dM_x}{dt} &= +\gamma[B_1 \sin \omega t M_z + B_0 M_y] - M_x/T_2\end{aligned} \tag{4-9}$$

These equations take a simpler form if they are referred to a set of axes (x', y', z) rotating with the applied field, with an angular velocity $-\omega$ about the z axis (Fig. 4-1). We shall call the components of magnetization in the plane perpendicular to B_0, u and v, where u is the component in the direction of B_1 along x' (the in-phase component), whilst v is the out-of-phase component along the y' axis.

$$\begin{aligned}u &= M_x \cos \omega t - M_y \sin \omega t \\ v &= +M_x \sin \omega t + M_y \cos \omega t\end{aligned} \tag{4-10}$$

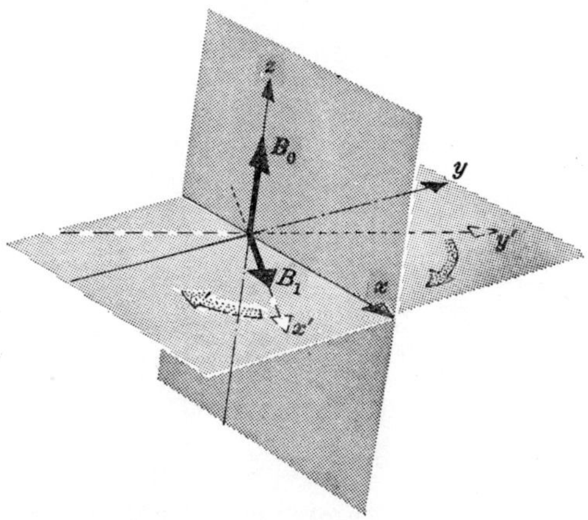

Fig. 4-1 System of rotating axes (x', y', z) using for solving the Bloch equations. These axes rotate about the z axis with constant angular velocity $-\omega$; in this system $\mathbf{B}_1$ is constant and is in the x' direction.

An NMR spectrometer can be set up to observe either u or v directly. The rate of change of u and v can be found from the Bloch equations [Eqn. (4-9)]

$$\frac{du}{dt} = \frac{dM_x}{dt}\cos\omega t - \frac{dM_y}{dt}\sin\omega t - \omega(M_x \sin\omega t + M_y \cos\omega t) \tag{4-11}$$

$$= +\gamma B_0 v - \frac{u}{T_2} - \omega v \tag{4-12}$$

$\frac{dv}{dt}$ can be found in a similar way. The complete set of Bloch equations in rotating axes is

$$\begin{aligned} \frac{dM_z}{dt} &= \gamma B_1 u - (M_z - M_0)/T_1 \\ \frac{du}{dt} &= +(\omega_i - \omega)v - u/T_2 \\ \frac{dv}{dt} &= -(\omega_i - \omega)u + \gamma B_1 M_z - v/T_2 \end{aligned} \tag{4-13}$$

where $\omega_i = \gamma B_0$.

In these axes the total magnetic field B_t is constant and equal to $(B_0^2 + B_i^2)^{1/2}$ (Fig. 4-2). However, because they are referred to rotating axes, these equations have extra terms (compared to the equations in static axes) which depend on ω, the angular velocity of the axes. Comparing Eqns. (4-4) and (4-13), we see that these terms have the

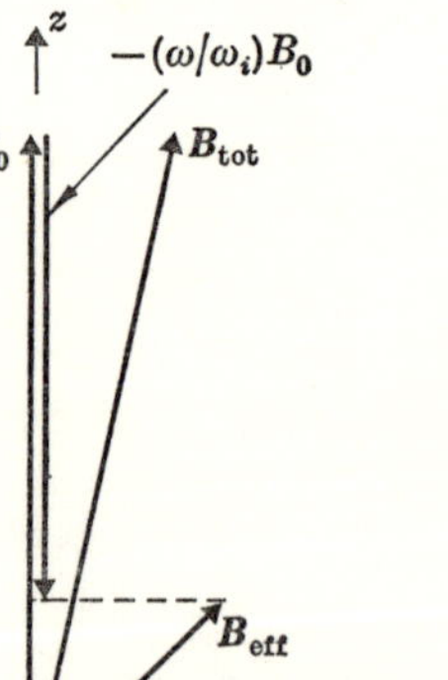

Fig. 4-2 Magnetic fields in the rotating axis system. As B_0, the static field, always much greater than the rotating field, B_1, the total field, $\mathbf{B}_{tot}$, is nearly parallel to $\mathbf{B}_0$. However the effective field $\mathbf{B}_{eff}$ may be in any direction in the zx' plane depending on the ratio ω/ω_i of the applied and Larmor angular frequencies.

same effect on the precessional motion as a magnetic field of $-\omega/\gamma$ in the z direction in stationary axes. In order to predict and understand the behaviour of the system, it is useful to define an effective field, $\mathbf{B}_{\text{eff}}$, with components $B_0\left(1 - \dfrac{\omega}{\omega_i}\right)$ in the z direction and B_1 in the x' direction. These are shown in Fig. 4-2. As the field B_1 is always much smaller than B_0, the total field $\mathbf{B}_t$ is nearly $\mathbf{B}_0$ and is close to the z direction. $\mathbf{B}_{\text{eff}}$, on the other hand, changes direction as the frequency ω of the radio frequency field changes.

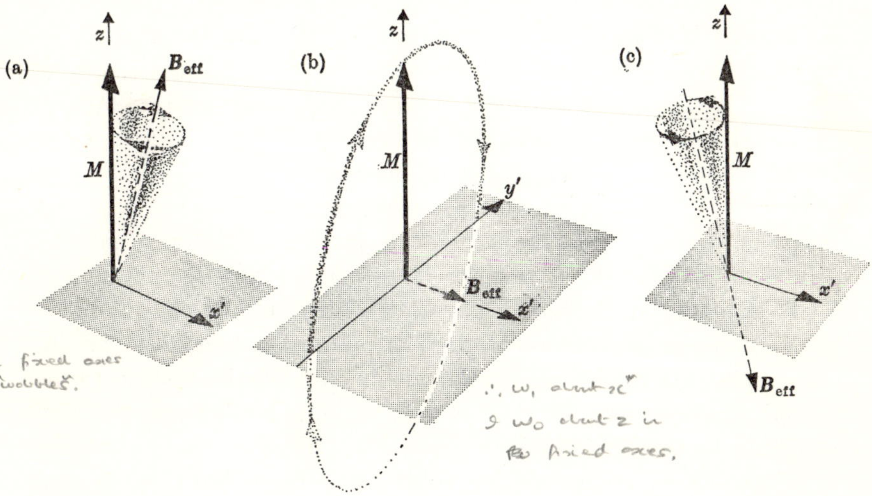

Fig. 4-3 Effective fields and the initial precession of the nuclear magnetization vector **M** in rotating axes with an applied radio frequency (a) below resonance (b) at resonance (c) above resonance. This precession is eventually damped by relaxation.

This is shown in the three diagrams in Fig. 4-3. In the first diagram ω is much less than ω_i, $\mathbf{B}_{\text{eff}}$ is nearly equal to $\mathbf{B}_0$ and, neglecting relaxation, the magnetization precesses about $\mathbf{B}_{\text{eff}}$ in a small cone. Diagram (b) shows the resonance situation where $\omega = \omega_1$ and $\mathbf{B}_{\text{eff}}$ is in the x' direction. Now **M** moves in a circle in the zy' plane, to give a large signal. In (c) $\omega \gg \omega_i$ and $\mathbf{B}_{\text{eff}}$ is nearly in the $-z$ direction. The behaviour of **M** is similar to the first case, except that the initial precession is in the opposite direction. It is only when ω is close to resonance that $\mathbf{B}_1$ has an appreciable effect on the system. When the field is first turned on, the magnetization precesses in the way described here, but after a short period it relaxes to a definite position at rest in the rotating axes. The signals observed in steady state experiments depend on the magnitudes of u and v when this has occurred. In pulse experiments the system is observed before the steady state is reached and the effects of T_1 and T_2 can be seen directly.

4-3 Pulse experiments

In a pulse experiment, the oscillating field is turned on for a time that is short compared to T_1 and T_2, and then turned off again. The frequency is chosen to be the resonant frequency, so that $\mathbf{B}_{\text{eff}} = \mathbf{B}_1$, as shown in Fig. 4-3(b). In rotating axes, the magnetization $\mathbf{M}$ precesses about the x' direction with the Larmor period of $2\pi/\gamma B_1$, so that a pulse lasting for $(n + \frac{1}{4})$ periods leaves $\mathbf{M}$ in the y' direction while one lasting for $(n + \frac{1}{2})$ periods reverses the direction of $\mathbf{M}$. These are known as 90° and 180° pulses respectively, and by measuring the rate of re-establishment of the equilibrium magnetization after such pulses, T_1 and T_2 can be measured.

There are two factors which contribute to the decay of v after a single 90° pulse. First, the Larmor frequencies of different portions of the sample differ slightly as a result of field inhomogeneities, and, in axes rotating at the mean Larmor frequency, the magnetization vectors of these different portions slowly fan out. Secondly, random processes tend to realign the nuclei within each portion to re-establish thermal equilibrium; this is true relaxation. Both these processes contribute to the line width in a steady state experiment, but they can be measured separately by applying a succession of pulses and observing spin echoes. The production of these echoes is shown in Fig. 4-4. An initial 90° pulse turns $\mathbf{M}$ into the y' direction in the system of rotating axes [Fig. 4-4(a)]. The magnetization vectors from different portions of the sample then begin to fan out (b) and the signal (proportional to v) decays. An 180° pulse is then applied after time t (c), which has the effect of rotating all the magnetization vectors by 180° about x', or in other words reflecting them in the $x'z$ plane. They continue to move in the same direction, and after a further time t, they are again in phase in the $-y'$ direction. The process can then be repeated by applying successive 180° pulses. The amplitudes of successive echos decay exponentially, due to relaxation, and the true value of T_2 can be found from this decay.

Pulse experiments can also be used to measure coupling constants, but the observations are more difficult to interpret than those of normal steady state experiments. The advantage is that they can be carried out with a magnet with poor field homogeneity, and the equipment is consequently much less expensive. The method has never become popular.

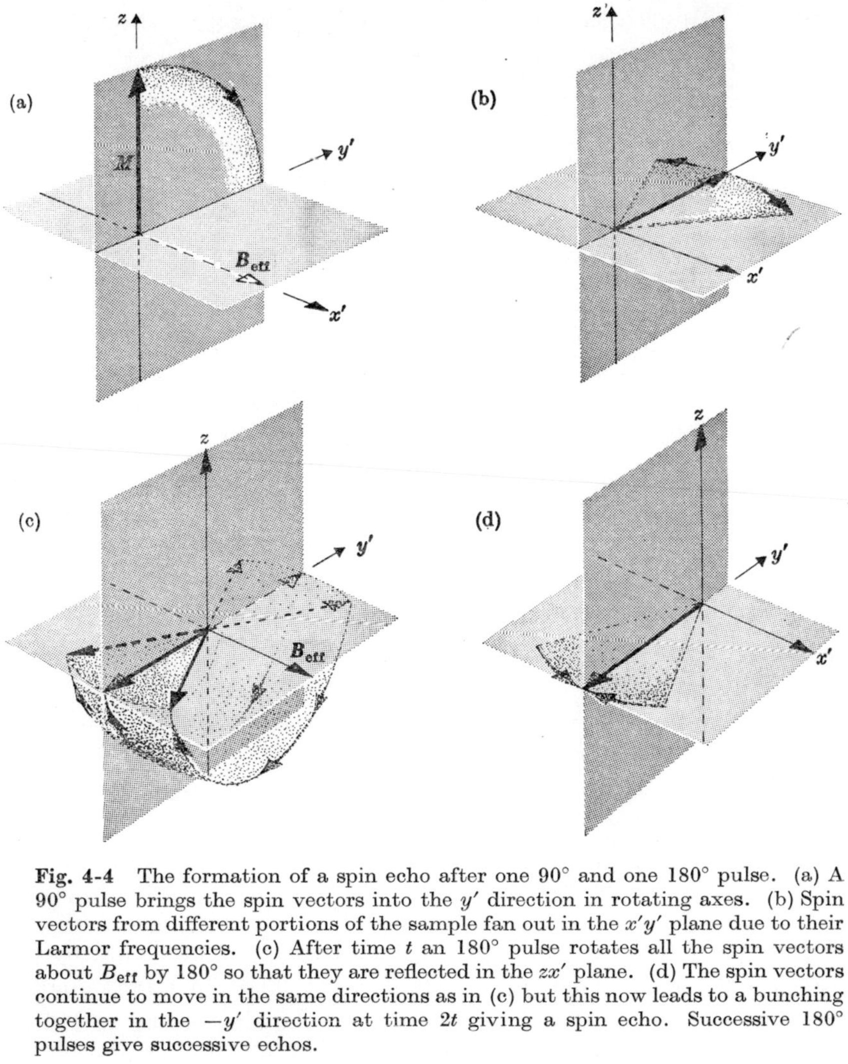

Fig. 4-4 The formation of a spin echo after one 90° and one 180° pulse. (a) A 90° pulse brings the spin vectors into the y' direction in rotating axes. (b) Spin vectors from different portions of the sample fan out in the $x'y'$ plane due to their Larmor frequencies. (c) After time t an 180° pulse rotates all the spin vectors about B_{eff} by 180° so that they are reflected in the zx' plane. (d) The spin vectors continue to move in the same directions as in (c) but this now leads to a bunching together in the $-y'$ direction at time $2t$ giving a spin echo. Successive 180° pulses give successive echos.

4-4 Steady state experiments

In a normal nuclear resonance experiment, the spectrometer is tuned to observe v, the component of magnetization 90° out of phase to the rotating field B_1. Occasionally u is observed, which gives the dispersion spectrum rather than the absorption spectrum. In a typical slow-passage experiment, ω is swept slowly through the resonance value and the signal at each frequency is recorded. Provided this is done sufficiently slowly, the signal at each frequency corresponds to the steady state value of ωv when $\mathbf{M}$ has come to rest in rotating axes. The values of u, v, and M_z are given by the solutions of the Bloch equations in rotating axes

[Eqn. (4-13)] with the time derivatives equal to zero. They are

$$M_z = \frac{M_0[1 + T_2^2(\omega_0 - \omega)^2]}{T_2^2(\omega_0 - \omega)^2 + 1 + T_1T_2\gamma^2B_1^2} \qquad (4\text{-}14)$$

$$u = \frac{M_0\gamma B_1 T_2^2(\omega_0 - \omega)}{T_2^2(\omega_0 - \omega)^2 + 1 + T_1T_2\gamma^2B_1} \qquad (4\text{-}15)$$

$$v = \frac{M_0\gamma B_1 T_2}{T_2^2(\omega_0 - \omega)^2 + 1 + T_1T_2\gamma^2B_1^2} \qquad (4\text{-}16)$$

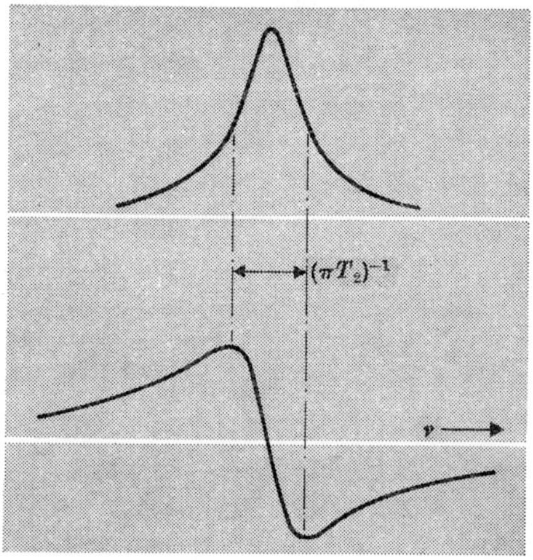

Fig 4-5 Lorentzian absorption and dispersion line shapes. The width of the absorption line at half-height and the peak-to-trough separation of the dispersion line are both equal to $(\pi T_2)^{-1}$.

Figure 4-5 shows the shape of the u and v signals predicted by the Bloch equations when $\gamma B_1 \ll (T_1T_2)^{-1/2}$. The v-mode signal is proportional to the power absorbed from the electromagnetic field, and is therefore analogous to infrared and ultraviolet absorption spectra, except that the NMR signal depends on B_1 and not on B_1^2. The u-mode signal gives the dispersion spectrum, which can be observed in other types of spectroscopy by measuring the refractive index of the sample as a function of frequency. This is only common in optical rotatory dispersion spectra, which depend on the difference of refractive indices of right- and left-handedly polarized light.

The line shapes predicted by the Bloch equations are known as Lorentzian and are the shapes that would be observed in an ideal experiment if the relaxation of M_x and M_y were truly exponential with a single value of T_2. In practice, the lines in nuclear resonance experiments are often not Lorentzian and may even be unsymmetrical. This is because the 'natural' width of the lines $(\pi T_2)^{-1}$ is so small that the observed width is due to instrumental effects, such as the variation of B_0 by a few parts per hundred million (0·6 Hz at 60 MHz) over the area of the sample. The line shape is an indication of the field inhomogeneity over the sample. This is sometimes known as inhomogeneity broadening, as the signal is a composite of lines with slightly different Larmor frequencies. The effects of field inhomogeneities can be minimized by spinning the sample.

4-5 Saturation

If the power in the oscillating field is too high, the intensity in the absorption spectrum decreases, owing to a reduction in the difference in population between the two levels. This occurs if the rate of energy absorption is comparable to, or greater than, the rate of relaxation between the energy levels, T_1^{-1}. The Bloch equations for a single proton show this effect when the term $\gamma^2 B_1^2 T_1 T_2$ in the denominator is no longer much less than one. Figure 4-6 shows the line shapes for various values of $\gamma^2 B_1^2 T_1 T_2$. As B_1 increases the lines get broader. The total intensity increases linearly with B_1 for low B_1, but tends to a constant value. Figure 4-7 shows the variation of peak height and

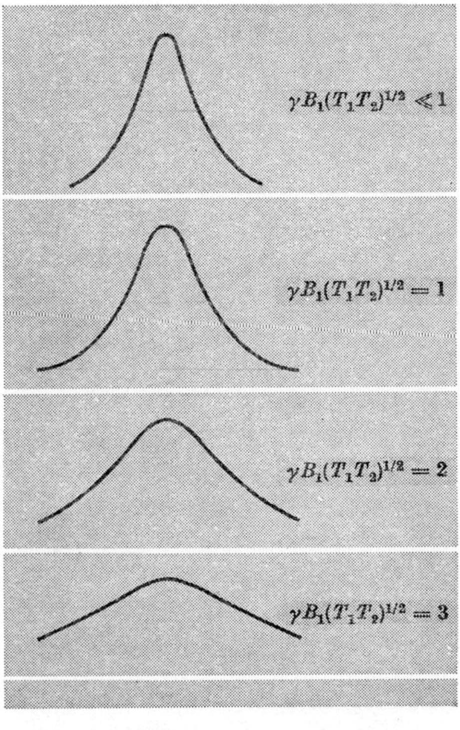

Fig. 4-6 Line shapes predicted by the Bloch equations under conditions of increasing saturation. The intensity of the unsaturated line is not to scale.

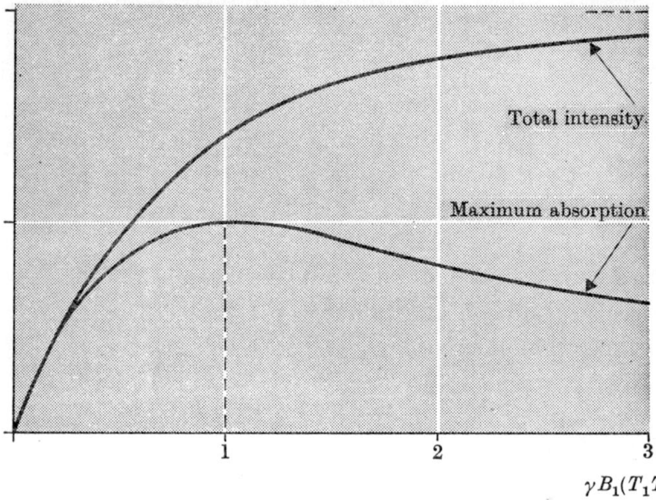

Fig. 4-7 Variation of peak height and total intensity with the amplitude of the observing field B_1.

total intensity as a function of B_1. From a practical point of view, a strong signal is best observed with $B_1 \ll (\gamma^2 T_1 T_2)^{-1/2}$, but if the concentration of spins is too low for this to be possible, the optimum conditions correspond to maximum peak height (although the signal will be distorted) with

$$B_1 = (\gamma^2 T_1 T_2)^{-1/2}$$

In a spectrum of a molecule with several lines with different relative intensities (such as those discussed in Chapter 2), the conditions for saturation vary from line to line. If the lines do not overlap and are not degenerate, each can be described by equations similar to the Bloch equations, but with different values of T_1 and T_2, and with B_1, u, and v replaced by fB_1, u/f and v/f, where f^2 is the intensity of the line, $|\langle m| \hat{F}_+ |n\rangle|^2$. The absorption line shape is proportional to v i.e. to

$$\frac{\gamma f^2 B_1 T_2 M_0}{T_2^2(\omega_0 - \omega)^2 + 1 + f^2 T_1 T_2 \gamma^2 B_1^2} \tag{4-17}$$

From an experimental point of view it should be noted that the size of the observing field B_1 can be increased to observe lines which have low transition probabilities, even if more intense lines in the same spectrum are saturated. However, lines which are weak because of concentration effects (such as ^{13}C satellites) cannot be enhanced in this way.

4-6 Causes of relaxation of spin $\frac{1}{2}$ nuclei

Relaxation between nuclear spin levels is much slower than relaxation of rotational or vibrational levels. The distortion of molecules during collisions often results in changes of rotational and vibrational states, but changing the orientation of a nucleus is a more difficult matter. Consequently, relaxation times of seconds are common in nuclear resonance, while rotational and vibrational relaxation times are more often of the order of 10^{-9} and 10^{-4} seconds respectively. A common temperature of vibrational, rotational and translational degrees of freedom is quickly established.

As the only property of a nucleus of spin $\frac{1}{2}$ which depends on orientation is its magnetic moment, transitions between nuclear spin levels can only be induced by magnetic fields. Thermal relaxation to give populations of the nuclear spin levels equal to those predicted by Boltzmann's distribution law with the same temperature as the bulk of the sample can only occur via random magnetic fields within the sample. A nucleus in a liquid will experience a fluctuating field, due to the magnetic moments of nuclei in other molecules as they execute Brownian motion. This random fluctuating field may be resolved by Fourier analysis into components oscillating at different frequencies. The component perpendicular to the static field which oscillates with the Larmor frequency induces transitions between the levels in a similar way to an electromagnetic field. This causes relaxation of both the longitudinal and transverse components of **M**. The populations of the states change until they reach the values predicted by the Boltzman equations for the temperature of the Brownian motion. This process is described by T_1, and also results in the relaxation of the longitudinal component of **M**. The contribution to T_2 is less obvious, but can be seen from the following argument. The line width, which is inversely proportional to T_2 (Fig. 4-5), is a measure of the uncertainty in the energies of the two states concerned. From the Heisenberg uncertainty principle this is inversely proportional to the lifetimes of these states, which we have seen are reduced by random fluctuations of the local magnetic field. Thus, fluctuations which cause transitions between states result in both the changes of population associated with longitudinal relaxation (T_1), and the increase in line width associated with transverse relaxation (T_2). In practice, the magnitudes of T_1 and T_2 are similar in fluids and in a number of cases are equal.

If there are paramagnetic species present (free radicals, oxygen or transition metal ions) relaxation is much more rapid, because the magnetic moment of an electron is so much greater than that of a nucleus. Intramolecular effects may also contribute to the establishment of thermal equilibrium between spin and rotational degrees of freedom, as the energy of a molecule that is static or in a definite rotational state has several orientation-dependent terms such as the anisotropic part of the magnetic shielding. Also, depending on the orientation (or rotational state) of the molecule, the nucleus will experience magnetic fields due to dipolar coupling with other nuclei within the molecule, and that due to the rotation of the molecule. All these average to zero under conditions of rapid tumbling and do not appear in the spin Hamiltonian, but as the molecule tumbles, they give fluctuating fields which may cause relaxation. In molecular beams, on the other hand, collisions are rare and these terms affect the positions of spectral lines.

4-7 Quadrupolar relaxation

Nuclei of spin greater than $\frac{1}{2}$ have electric quadrupole moments which cause rapid relaxation both of these nuclei and also, in some circumstances, of spin $\frac{1}{2}$ nuclei in the same molecule.

Electric multipole moments are due to non-spherical charge distributions, such as those illustrated in Fig. 4-8. These may result in orientation dependent energies. Thus a body with an electric dipole moment is aligned by an electric field, a body with an electric quadrupole moment interacts with the gradient of an electric field, a body with an electric octupole moment interacts with the second derivative of an electric field and so on. The quadrupole moment of a nucleus is characterized by a single parameter Q with dimensions (length)^2. In an axially symmetrical environment (such as a linear molecule), the energy of the interaction of the quadrupole moment

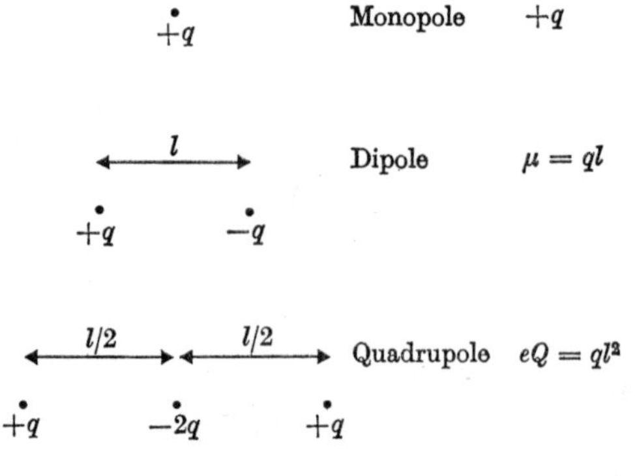

Fig. 4-8 Distribution of point charges with electric multipole moments. An arbitrary distribution of charges can always be resolved into a sum of multipole moments.

and the electric potential is described by a term in the spin Hamiltonian

$$\hat{\mathscr{H}}_Q = \frac{e^2Qq}{16\pi\epsilon_0 I(2I - 1)}(3\hat{I}_z^2 - \hat{\mathbf{I}}^2) \tag{4-18}$$

where $\epsilon_0\Xi = c^2$, $\hat{I}_z$ is the spin operator in the direction of the axis of symmetry of the environment (*not* the direction of the external field), and q is the electric field gradient $\dfrac{d^2V}{dz^2}$ in the z direction.

The electric field in a molecule is due to the electrons and nuclei. It has a value of zero at the equilibrium position of each nucleus, but unless the nucleus is in a position of high symmetry (for example the central atom in a tetrahedral ion such as NH_4^+) there is a finite field gradient. In such an environment, nuclear spin states have different energies, even in the absence of a magnetic field (Problem 4-4). Transitions between these states are observed in nuclear quadrupole spectroscopy, (see Suggestions for Further Reading). Interpretation of these spectra gives the value of q, the local field gradient.

In the nuclear resonance experiment, the interaction of the quadrupole moment with the molecular electric field relaxes the Zeeman states split by the applied magnetic field. In the absence of an external field, the orientation of the nucleus *relative to the molecule* is constant as the molecule tumbles. A spin $\frac{1}{2}$ nucleus in an applied field keeps a constant orientation *relative to the external field.* If the nucleus with a quadrupole moment is placed in an external field, there are two opposing influences, one trying to preserve the orientation relative to the laboratory, and the other relative to the molecule. For some nuclei, such as ^{14}N, ^{2}D, ^{11}B, ^{10}B, the interaction with the applied field is much greater than the quadrupolar interaction. The states are determined by the Zeeman splitting, but the quadrupole interaction causes relaxation.* Consequently, nuclear resonance can normally be observed, but the lines are much broader than for spin $\frac{1}{2}$. The breadth reflects the value of the electric field gradient. For example, signals from quaternary nitrogen atoms are sharper than from tertiary ones.

* And also removes the degeneracy of the Zeeman split levels.

Other nuclei, such as the halogens, have larger quadrupole interactions, either because their quadrupole moments are larger, or because they commonly occur in positions with higher field gradients. Pure quadrupole spectra of ^{35}Cl are usually observed between 10 and 30 MHz, while the splitting of the levels due to a magnetic field in which proton levels are split by 60 MHz is only 6 MHz. The spectral lines are too broad to be detected.

Effects of quadrupolar relaxation are also observed in the spectra of spin $\frac{1}{2}$ nuclei coupled to nuclei with quadrupole moments. This is illustrated by the spectrum of a molecule containing H and D. In the absence of quadrupolar relaxation, the proton spectrum [Fig. 4-9(a)] would consist of three sharp lines of equal intensity corresponding to m values of 1, 0, and -1 for the deuterium nucleus.

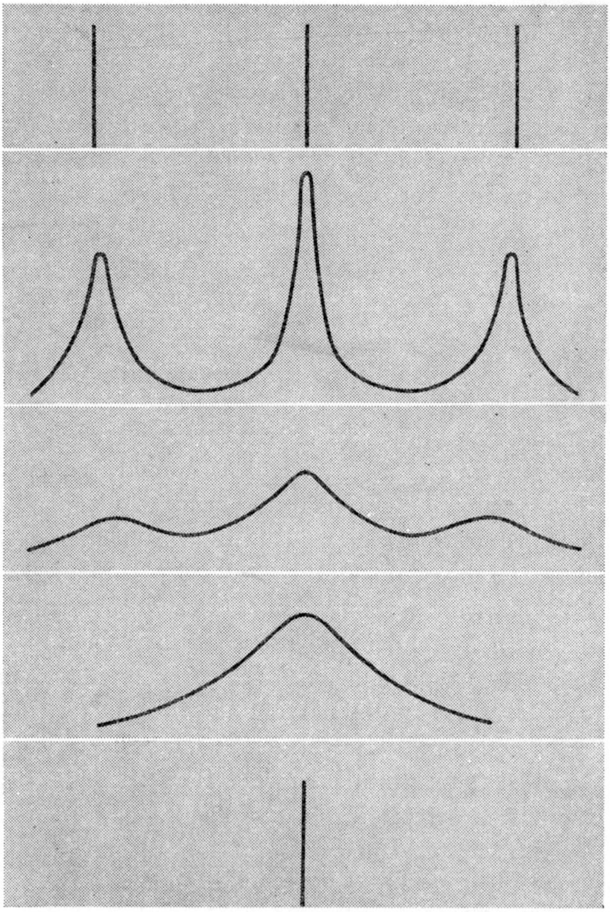

Fig. 4-9 Proton spectra of a molecule containing a proton coupled to a deuterium or ^{14}N nucleus showing the changes as the rate of quadrupolar relaxation increases. Note that the outer lines broaden more rapidly than the central lines. If the lines were broadened by chemical exchange the lines would initially have the same width.

Owing to quadrupolar relaxation, the states of the deuterium nucleus have rather short lifetimes τ_m. From the Heisenberg uncertainty principle, a state with a lifetime τ_m has an uncertainty in its energy of ΔE, where

$$\Delta E = h(2\pi\tau_m)^{-1} \tag{4-19}$$

If this is less than the coupling of the nuclei, hJ_{HD}, then the lines are broadened [Fig. 4-6(b)] by $(2\pi\tau_m)^{-1}$. If relaxation between the deuterium levels is very rapid compared with J, the proton spectrum no longer shows coupling with the deuterium nucleus, but becomes a single sharp line. At intermediate rates of relaxation, intermediate spectra are seen [Fig. 4-9(c) and (d)].

Deuterium substitution is sometimes used to determine the coupling between magnetically equivalent protons. Figure 4-10

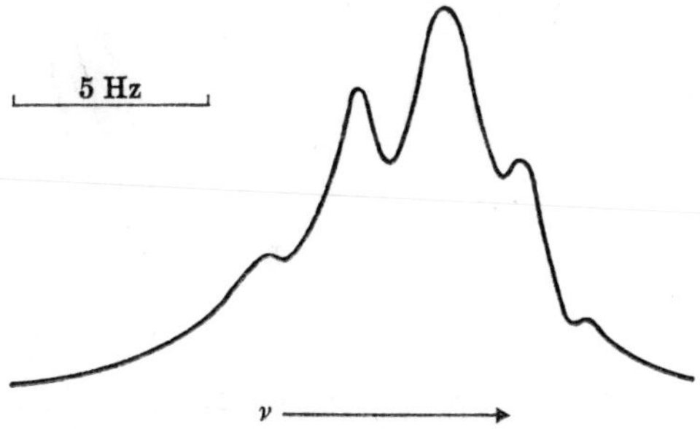

Fig. 4-10 One component of the phosphorus doublet in the proton spectrum of PHD_2, showing the fine structure of five lines due to HD coupling. These are broadened by the quadrupolar relaxation of the deuterium nuclei. From the value of J_{HD} it can be deduced that J_{HH} in PH_3 is 13 Hz (reproduced from *Trans. Faraday Soc.*, 1961, **57,** 888).

shows part of the proton spectrum of PHD_2. The complete proton spectrum consists of a doublet with separation J_{PH}, each component of which is split into a 1:2:3:2:1 quintet with separation J_{HD}. Although these lines are broadened due to deuterium quadrupolar relaxation, J_{HD} can still be found and J_{HH} for PH_3 deduced by multiplying by γ_H/γ_D (see Section 3-16).

The rates of quadrupolar relaxation of chlorine, bromine, or iodine are always large compared to their coupling with protons or fluorine, so the spectra of the latter nuclei do not show evidences of coupling to the halogens. For the purposes of high-resolution nuclear resonance, chlorine, bromine, and iodine may be regarded as being non-magnetic.

4-8 Chemical exchange

One of the important chemical uses of nuclear resonance is the study of rates of reaction under equilibrium conditions. The range of rate constants for which this is possible is fairly small, but includes those for many conformational changes, as well as some solute–solvent interactions. The term 'chemical exchange' is used to describe reactions at equilibrium. The total numbers of nuclei in each environment remain constant, but each individual nucleus changes between different environments. It is convenient to distinguish between inter- and intra-molecular exchange processes. In an intramolecular process, the nuclei of each molecule remain bonded and coupled to each other, but when the molecule changes conformation the magnetic environment of each nucleus changes. An example is the inversion of cyclohexane. In solution, cyclohexane exists in the chair conformation with two distinct chemical environments for the protons, axial and equatorial. This conformation is not fixed but

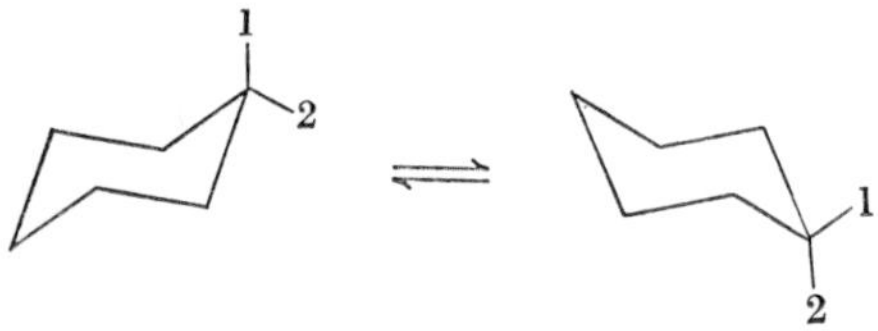

can invert into an equivalent chair form. As a result of this inversion, every axial proton becomes an equatorial proton and vice versa, but as the exchange is intramolecular, proton 1 remains coupled to 2.

In an intermolecular process, the molecules break into fragments which recombine with different fragments. As two nuclei originally in the same molecule have a negligible probability of recombining, the coupling is broken. Methanol in the presence of traces of acid shows intermolecular exchange. The hydroxyl protons jump rapidly from one molecule to another via a positively charged intermediate.

$$CH_3O_aH + CH_3O_bH_2^+ \rightleftharpoons CH_3O_aH_2^+ + CH_3O_bH$$

As the nuclei move from one molecule to another the coupling is broken.

The effect of such exchange processes on the nuclear resonance spectrum depends mainly on the rate at which they occur. If the nucleus is exchanging very slowly (slow exchange limit), the spectrum is a superposition of the spectra of the various species. If exchange is very fast, the spectrum depends only on the mean magnetic environment of the nuclei (fast exchange limit). Intermediate rates of exchange give spectra with broad lines from which the rate can be determined.

This change in the NMR spectrum as the rate of exchange is increased is illustrated by a system containing protons which jump between two sites A and B. There is no coupling, but there is a chemical shift between the two sites, which are assumed to have equal populations. Figure 4-11 shows the line shapes observed for various values of k, the rate at which a nucleus leaves one site and arrives at the other. As the two sites are equally populated, the rate constants for the forward and backward reactions are equal.

In the limit of zero exchange, the spectrum consists of two sharp lines at ν_A and ν_B. If the rate is small, the lines are broadened. This

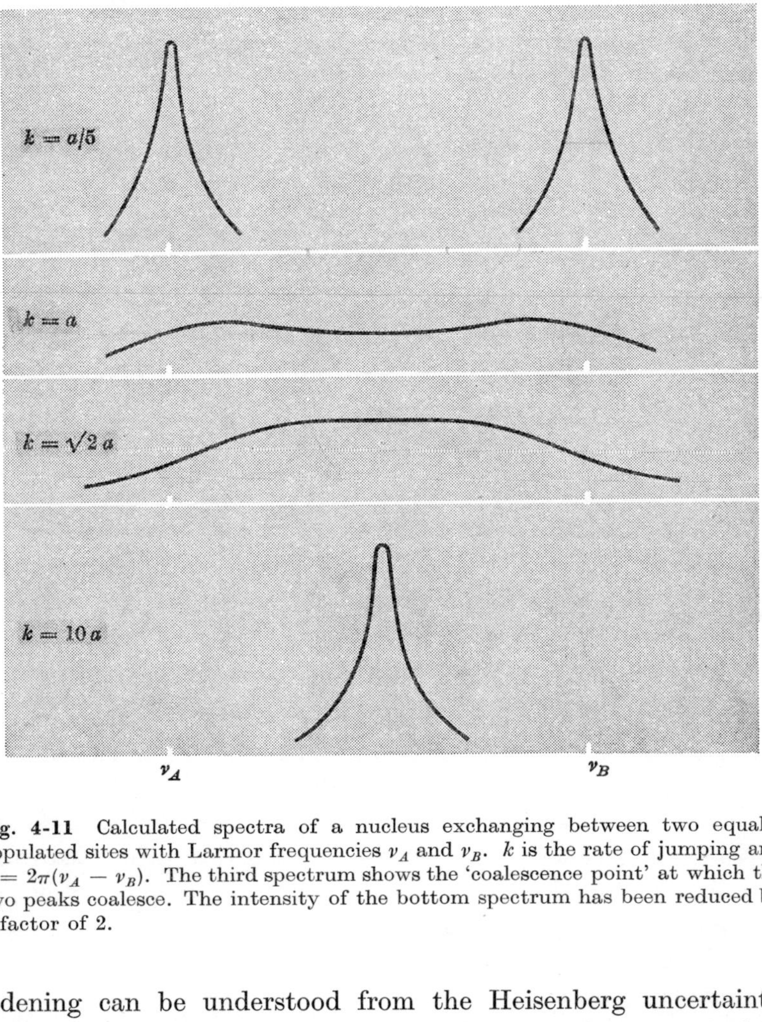

Fig. 4-11 Calculated spectra of a nucleus exchanging between two equally populated sites with Larmor frequencies ν_A and ν_B. k is the rate of jumping and $a = 2\pi(\nu_A - \nu_B)$. The third spectrum shows the 'coalescence point' at which the two peaks coalesce. The intensity of the bottom spectrum has been reduced by a factor of 2.

broadening can be understood from the Heisenberg uncertainty principle. The average lifetime of a nucleus in either state is k^{-1}, which gives an uncertainty in the energy of $\Delta E = \hbar k$ or $k/2\pi$ in the frequency. This broadening is analogous to the broadening by relaxation considered in the previous section with $T_2 = k^{-1}$, so each line is Lorentzian (or approximately so) with a width at half-height of k/π. When the lines begin to overlap appreciably, the line shape is no longer the sum of two Lorentzians. The maxima approach each other and there is a build-up of intensity in the centre. The line widths are no longer related to k in a simple way. At the other extreme of very rapid exchange, there is a single line at the average frequency of the two sites, with width $\frac{1}{4}(\nu_A - \nu_B)^2(\pi k)^{-1}$.

It is possible to find k from direct measurement of line widths in the slow exchange limit if $k \geqslant \frac{1}{2}\pi\ s^{-1}$. The line widths must be corrected for instrumental broadening and for natural width. Normally, the latter is so small that the correction can be made by subtracting the line width of an internal standard. In the fast exchange, limit k can only be found if ν_A and ν_B are known. In the intermediate region, k can be found by matching the line shape with calculated line shapes if ν_A and ν_B are known. Nuclear resonance can only be used to measure rates of processes whose first-order rate constants lie between approximately $\frac{1}{2}\ s^{-1}$ and $\Delta^2/2\pi\ s^{-1}$, where Δ is the spread of the spectrum in the absence of exchange, measured in hertz. However, it is possible to vary the temperature of the sample to bring the rate constants within this range.

In conformational studies, two quantities are of interest: the relative amounts of the different conformers and the barrier to internal conversion between conformers. The difference in energy of the conformers can be found by observing how the first of these varies with temperature. This is easily done using NMR methods, provided that interconversion is so slow that separate signals are seen from the different conformers. The determination of a barrier height is illustrated in Fig. 4-12 for 4,4-dimethylaminopyrimidine. Here there are two equivalent conformers with equal populations. In the last spectrum of Fig. 4-12, two peaks are seen in the methyl region of the spectrum, corresponding to the two non-equivalent methyl groups. The chemical shift difference between the groups *cis* and *trans* to the ring nitrogen is determined from this to be 17·8 Hz. As the temperature is increased, the rate of rotation about the C—N bond increases, until at room temperature (not shown) the spectrum is a single sharp line. In order to determine the height of the barrier to rotation about this bond, the rate constant k must be found at each temperature. k is related to the barrier height by the Arrhenius equation

$$k = A \exp\,(-E_a^{+}/RT) \qquad (4\text{-}20)$$

By plotting $\log_e k$ against $1/T$, the barrier height is found to be $4{\cdot}49 \times 10^4$ J mole^{-1} (11·3 kcal/mole) for a $CHCl_3$ solution.

Many other problems of conformation have been tackled by means of nuclear resonance, which has proved to be an invaluable tool in this field. For further details the reader is referred to relevant textbooks and reviews.

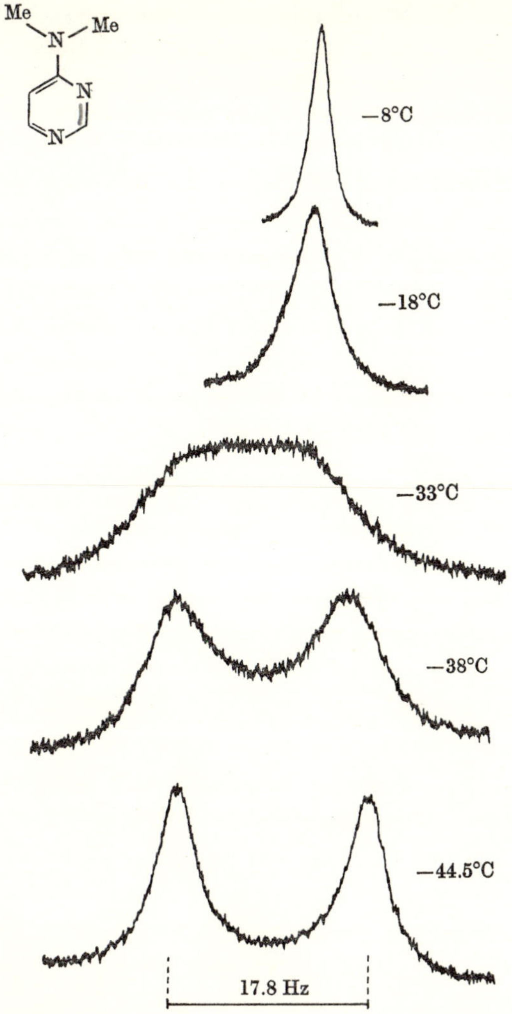

Fig. 4-12 Spectra of the methyl groups of 4,4-dimethylaminopyrimidine at 100 MHz and various temperatures. This series shows a sequence of spectra essentially of the form of Fig. 4-11, as there is no appreciable coupling between the methyl protons and any other nuclei. (We are grateful to A. R. Katritzky and G. J. T. Tiddy for this spectrum.)

PART B DOUBLE QUANTUM AND DOUBLE RESONANCE SPECTRA

4-9 Two-quantum transitions

The interaction of electromagnetic radiation with matter can lead to many effects, depending on the frequency and intensities of the radiation and on the wave function describing the matter. It is convenient to classify these effects according to the number of quanta involved. Single quantum processes include emission and absorption and all these can be considered as examples of a generalized scattering process. For example, in absorption the molecule in state M meets a quantum of frequency $h\nu$ and, after interaction, emerges in state M'. The initial and final molecular states are eigenstates of

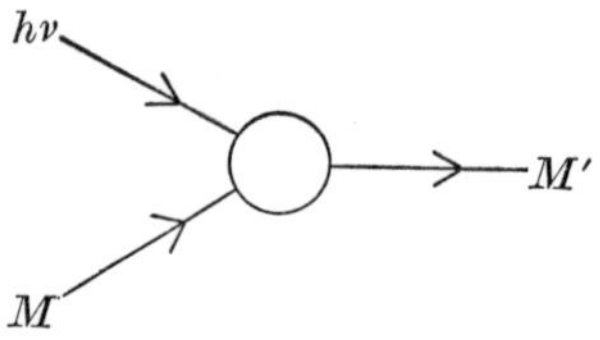

One-quantum absorption

the molecular Hamiltonian in the absence of the electromagnetic field. Conservation of energy leads to $E_M - E'_M = h\nu$. This is a one-quantum or single quantum effect because one quantum of electromagnetic radiation is involved. The best known example of a two-quantum process is Raman scattering. Other examples of

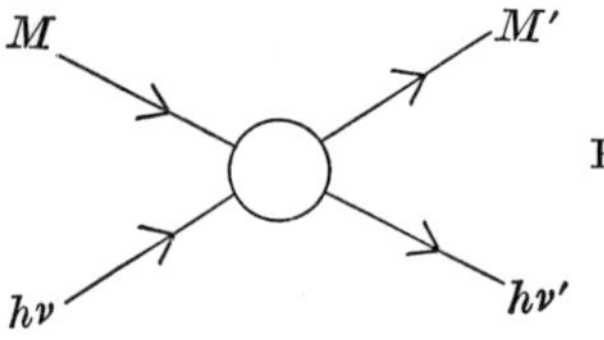

Raman Process

two-quantum processes have been observed using lasers and in nuclear resonance. The radio frequency field in a nuclear resonance experiment is in some ways similar to a laser beam, as it is coherent and may have a high power, especially if the latter is measured in quanta per second.

The two-quantum process often observed in nuclear resonance is the simultaneous absorption of two quanta of the same frequency.

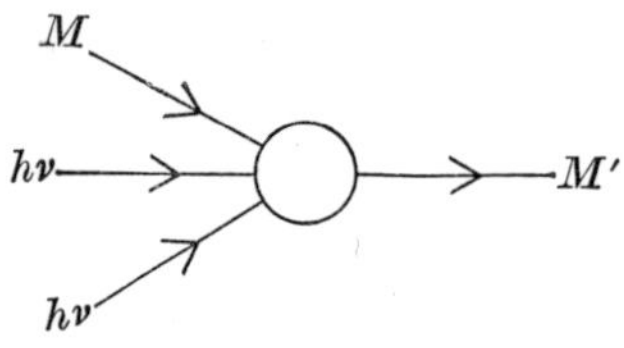

Two-quantum absorption

Conservation of energy leads us to the frequency condition

$$E_{M'} - E_M = 2h\nu \quad (4\text{-}21)$$

The selection rule for two-quantum transitions of this type is

$$\Delta m_T = \mp 2 \quad (4\text{-}22)$$

Thus, an AX system has a single two-quantum transition $\alpha\alpha \rightarrow \beta\beta$ at $\frac{1}{2}(\nu_A + \nu_X)$.

The detailed theory of this effect shows that the intensity is small unless the two-quantum transition is fairly close to the ordinary absorption lines. Essentially, an AB system is required. The intensity of the lines depends on B_1^3 (B_1 for single quantum).

Two-quantum transitions are sometimes useful in analysing spectra, and, in particular, in determining the relative signs of coupling constants.

4-10 Double resonance

In a double resonance experiment two radio frequency fields are applied simultaneously to the sample. Several different phenomena can be observed, depending on the frequencies and amplitudes of these oscillating fields, ranging from changes in intensity to completely altered spectra. The first radio frequency field $\mathbf{B}_1 \cos \omega_1 t$ is known as the observing field and is analogous to the radio frequency field in an ordinary single resonance experiment. The second, or perturbing, radio frequency field, $\mathbf{B}_2 \cos \omega_2 t$, is also applied perpendicular to $\mathbf{B}_0$, the static field. The effects of $B_2 \cos \omega_2 t$ on the spin system are observed using the field $B_1 \cos \omega_1 t$ to monitor the signal at angular frequency ω_1. In a single resonance experiment, identical spectra can be obtained by field sweep (changing B_0) and frequency sweep (changing ω_1) methods (see Section 1-16). This is no longer true for double resonance, as changing B_0 is equivalent to simultaneously changing both ω_1 and ω_2. It is usual to use the frequency sweep method, so that a double resonance spectrum corresponds to a fixed value of B_0 and fixed amplitude and frequency of the perturbing field. Different double resonance spectra are obtained by changing the parameters of the perturbing field.

If the amplitude of the perturbing field B_2 is small and its frequency corresponds to a transition of the spin system, the double resonance spectrum will have lines at the same frequencies as the single resonance spectrum, but will show changes of intensity. This is illustrated in Fig. 4-13(a), where P, Q, and R are nuclear spin energy levels with allowed transitions from R to Q and Q to P. If the transition from R to Q is saturated by the field B_2, then the population of Q is increased and that of R decreased. As the intensity of the transition Q to P depends on the difference in populations of levels Q and P, it is enhanced by the increase in the population of Q. Other transitions involving one of the perturbed levels have enhanced or lowered intensities. The detailed study of these intensity changes can give information about relaxation processes. Such effects are known as Overhauser effects.

Two-quantum transitions can be observed when ω_2 does not

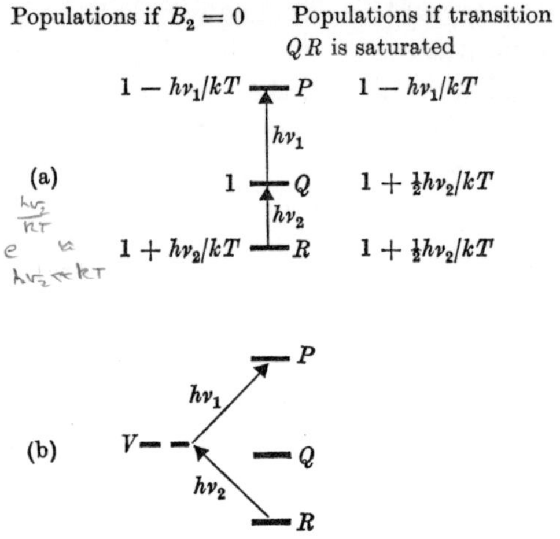

Fig. 4-13 Overhauser and double quantum transitions. (a) If the perturbing field is relatively weak and applied at resonance it changes the populations of the levels Q, R. The intensity of transition QP is enhanced. (b) If ν_2 is off resonance a two-quantum transition occurs when $(h\nu_1 + h\nu_2)$ equals the difference in energy of levels P and R.

correspond exactly to a transition. This is shown in the energy level diagram 4-13(b). The two-quantum transition is a transition from state R to state P in which two quanta are absorbed, one from B_1 and one from B_2. The double resonance spectrum in this case will contain all the single resonance lines and in addition these two-quantum transitions. The intensity of the latter are small, and depend both on the value of $B_2^2B_1$ and on the difference in energy between the virtual level V and the state Q.

Most double resonance experiments are of neither of these two types, but employ a much larger perturbing field so that the spin energy levels themselves are altered. Similar effects have been observed with lasers in the optical region, but in the radio region it is easier to obtain electromagnetic radiation with sufficient amplitude and coherence to perturb molecular energy levels. In these circumstances it is no longer possible to describe the spectrum in terms of one- or two-quantum transitions between the energy levels of an isolated molecule in a field B_0; instead, we must determine the energy levels of the spin system in the presence of two fields, B_0 and $B_2 \cos \omega_2 t$, one of which is time dependent. In Section 4-11 this problem is discussed theoretically, using similar techniques to those described in Chapter 2. The succeeding sections describe the types of observations made and discuss their applications with examples. These sections should be comprehensible by the reader who does not wish to examine the theoretical detail of Section 4-11.

4-11 Theoretical description of double resonance

Double resonance spectra which are more complex than the simple two-quantum transitions described in the previous section can most easily be interpreted as a one-quantum spectrum observed with the measuring field $B_1 \cos \omega_1 t$, in which transitions occur between the levels of the combined system of spins and fields B_0 and $B_2 \cos \omega_2 t$. The problem is to determine the energy levels and wave functions of the latter system. In order to do this, we first remove the time dependence of the perturbing field by using axes which rotate with the latter field.

In Section 4-2, transformation to rotating axes was discussed for a classical system and it was found that the equations of motion in axes rotating about the z direction with angular frequency $-\omega$ contain extra terms which have the same form as an additional magnetic field of $-\omega\gamma^{-1}$ in the z direction. Similarly we can write an effective Hamiltonian in rotating axes (x, y, z) (see Problem 4-7)

$$\hat{\mathscr{H}}_{\text{eff}} = \sum_j -(2\pi)^{-1}\gamma_j\{[(1-\sigma_j)B_0 - \omega_2\gamma_j^{-1}]\hat{I}_{jz} + B_2\hat{I}_{jx}\} + \sum_{j<k} J_{jk}\hat{\mathbf{I}}_j \cdot \hat{\mathbf{I}}_k \quad (4\text{-}23)$$

whose eigenstates are stationary in the system of rotating axes. This Hamiltonian differs from that given in Chapter 2 in two respects. First, there is a change from B_0 to $(B_0 - \omega_2\gamma_j^{-1})$, and secondly there is an additional term involving the spin operators $\hat{I}_{jx}$. This latter term mixes states with different values of m, which is no longer a good quantum number.

The method of finding the eigenstates and eigenvalues of this double resonance Hamiltonian is to set up the secular equations and to solve them by diagonalizing the Hamiltonian matrix as described in Section 2-9.

The AX system

The spin Hamiltonian for the AX system in rotating axes is

$$\hat{\mathscr{H}}_{\text{eff}} = -(\nu_A - \nu_2)\hat{I}_{Az} - (\nu_X - \nu_2)\hat{I}_{Xz} + (2\pi)^{-1}B_2(\gamma_X\hat{I}_{Xx} + \gamma_A\hat{I}_{Ax}) + J\hat{\mathbf{I}}_A \cdot \hat{\mathbf{I}}_X \quad (4\text{-}24)$$

where $\nu_A = \gamma_A(1-\sigma_A)B_0$ and $2\pi\nu_2 = \omega_2$.

If ν_2 is in the X region, we may omit the term in $B_2\hat{I}_{Ax}$ as the latter mixes states with m_A values which differ by unity. As the difference in diagonal elements of such states $(\nu_A - \nu_2)$ is large compared with $\gamma_A B_2/2\pi$, the term in $\hat{I}_{Ax}$ has a negligible effect. Similarly, we may make the X approximation and omit the terms $\frac{1}{2}J\hat{I}_{A+}\hat{I}_{X-}$ and

$\frac{1}{2}J\hat{I}_{A-}\hat{I}_{X+}$ (see Section 2-10), leaving an effective Hamiltonian

$$\hat{\mathscr{H}}_{\text{eff}} = -(\nu_A - \nu_2)\hat{I}_{Az} - (\nu_X - \nu_2)\hat{I}_{Xz} + (2\pi)^{-1}B_2\gamma_X\hat{I}_{Xx} + J\hat{I}_{Az}\hat{I}_{Xz} \tag{4-25}$$

for which the value of m_A is a good quantum number as $\hat{\mathscr{H}}_{\text{eff}}$ commutes with $\hat{I}_{Az}$. As the effective field for the A nuclei (Fig. 4-14) is still nearly in the z direction, they are quantized, while the X nuclei are not.

This Hamiltonian, like the normal AX Hamiltonian, has four eigenvalues. However, m_T is no longer a good quantum number and

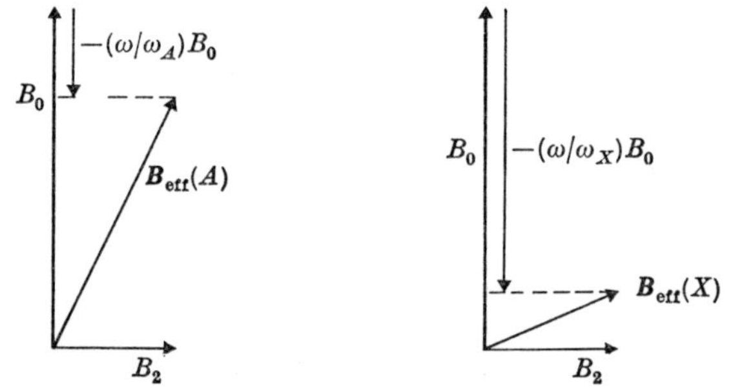

A X

Fig. 4-14 Effective fields in rotating axes for A and X nuclei in an $A-\{X\}$ double resonance experiment.

there are now six possible transitions, two in the X and four in the A region. The frequencies and intensities of these transitions depend on the values of ν_2 and B_2. Referred to the set of states $(\alpha\alpha,\ \alpha\beta,\ \beta\alpha,\ \beta\beta)$ defined in the rotating frame, the Hamiltonian matrix is

$$\begin{bmatrix} \frac{1}{2}(2\nu_2-\nu_A-\nu_X)+\frac{1}{4}J & \gamma_X B_2/4\pi & 0 & 0 \\ \gamma_X B_2/4\pi & \frac{1}{2}(\nu_X-\nu_A)-\frac{1}{4}J & 0 & 0 \\ 0 & 0 & \frac{1}{2}(\nu_A-\nu_X)-\frac{1}{4}J & \gamma_X B_2/4\pi \\ 0 & 0 & \gamma_X B_2/4\pi & \frac{1}{2}(\nu_X+\nu_X-2\nu_2)+\frac{1}{4}J \end{bmatrix} \tag{4-26}$$

Table 4-1 Eigenvalues and eigenfunctions of the double resonance Hamiltonian for the AX system

State[a]	*Eigenvalue*
$\cos\theta_+ \lvert\alpha\alpha\rangle + \sin\theta_+ \lvert\alpha\beta\rangle$	$\frac{1}{2}(\nu_2 - \nu_A) + \frac{1}{2}[(\nu_2 - \nu_X + \frac{1}{2}J)^2 + (\gamma_X B_2/2\pi)^2]^{1/2}$
$-\sin\theta_+ \lvert\alpha\alpha\rangle + \cos\theta_+ \lvert\alpha\beta\rangle$	$\frac{1}{2}(\nu_2 - \nu_A) - \frac{1}{2}[(\nu_2 - \nu_X + \frac{1}{2}J)^2 + (\gamma_X B_2/2\pi)^2]^{1/2}$
$\cos\theta_- \lvert\beta\alpha\rangle + \sin\theta_- \lvert\beta\beta\rangle$	$-\frac{1}{2}(\nu_2 - \nu_A) + \frac{1}{2}[(\nu_2 - \nu_X - \frac{1}{2}J)^2 + (\gamma_X B_2/2\pi)^2]^{1/2}$
$-\sin\theta_- \lvert\beta\alpha\rangle + \cos\theta_- \lvert\beta\beta\rangle$	$-\frac{1}{2}(\nu_2 - \nu_A) - \frac{1}{2}[(\nu_2 - \nu_X - \frac{1}{2}J)^2 + (\gamma_X B_2/2\pi)^2]^{1/2}$

[a] $\sin 2\theta_+ = \gamma_X B_2/2\pi[(\nu_2 - \nu_X + \frac{1}{2}J)^2 + (\gamma_X B_2/2\pi)^2]^{1/2}$,
$\sin 2\theta_- = \gamma_X B_2/2\pi[(\nu_2 - \nu_X - \frac{1}{2}J)^2 + (\gamma_X B_2/2\pi)^2]^{1/2}$.

The eigenvalues and eigenfunctions are given in Table 4-1. If the double resonance spectrum is observed in the A region, transitions in which $\Delta m_A = -1$ are seen. There are four such transitions whose frequencies can be determined from Table 4-1. However, it must be remembered that this table refers to rotating axes. An applied field with frequency ν_1 in laboratory axes appears to have frequency $\nu_1 - \nu_2$ in rotating axes, so the appropriate resonance condition for transition between states $|m\rangle$ and $|n\rangle$ is

$$\nu_1 - \nu_2 = \mathscr{H}_{mm} - \mathscr{H}_{nn} \tag{4-27}$$

The relative intensities depend on

$$|\langle m| \sum \gamma_j \hat{I}_{j-} |n\rangle|^2 \tag{4-28}$$

where $\hat{I}_-$ acts in the rotating frame.

4-12 Tickling experiments

If one particular transition of a multispin system is irradiated by the perturbing field, the two levels connected by that transition are perturbed, and other transitions involving these levels will be the first to be affected. We have already seen that the intensities of these transitions are altered by relatively weak perturbing fields. As the strength of the perturbing field is increased, these transitions are split in the double resonance spectrum by an amount $\gamma B_2/2\pi$. This allows one to identify transitions which share a common energy level, and so to assign lines in a complex spectrum and in many cases to find the relative signs of coupling constants. Such an experiment in which one transition is selectively irradiated and the resulting line splittings are observed is known as a tickling experiment.

Figure 4-15 shows the double resonance spectrum of the M and A parts of an AMX system in which the lowest frequency line of the X group is irradiated. The molecule is $CCl_3CHClCH_2Cl$, in which the two methylene protons are non-equivalent because of restricted internal rotation about the C—C bond and intrinsic asymmetry.

Figure 4-16 shows the energy level diagram and the single resonance spectrum of this molecule. The lines are labelled to correspond to the transitions on the energy level diagram for four possible choices of signs of coupling constants In the double resonance spectrum shown in Fig. 4-15, the line marked with an arrow is being tickled, and the lines marked with stars are observed to be affected. From the energy level diagram, we see that irradiating any X transition will affect two A transitions; if X_2 is tickled, A_3 and A_4 are affected, while tickling X_4 affects A_3 and A_4. The third assignment is consistent with the observations, while the others are ruled out. The reader should ascertain for himself that it can be deduced

WADHAM COLLEGE
LIBRARY

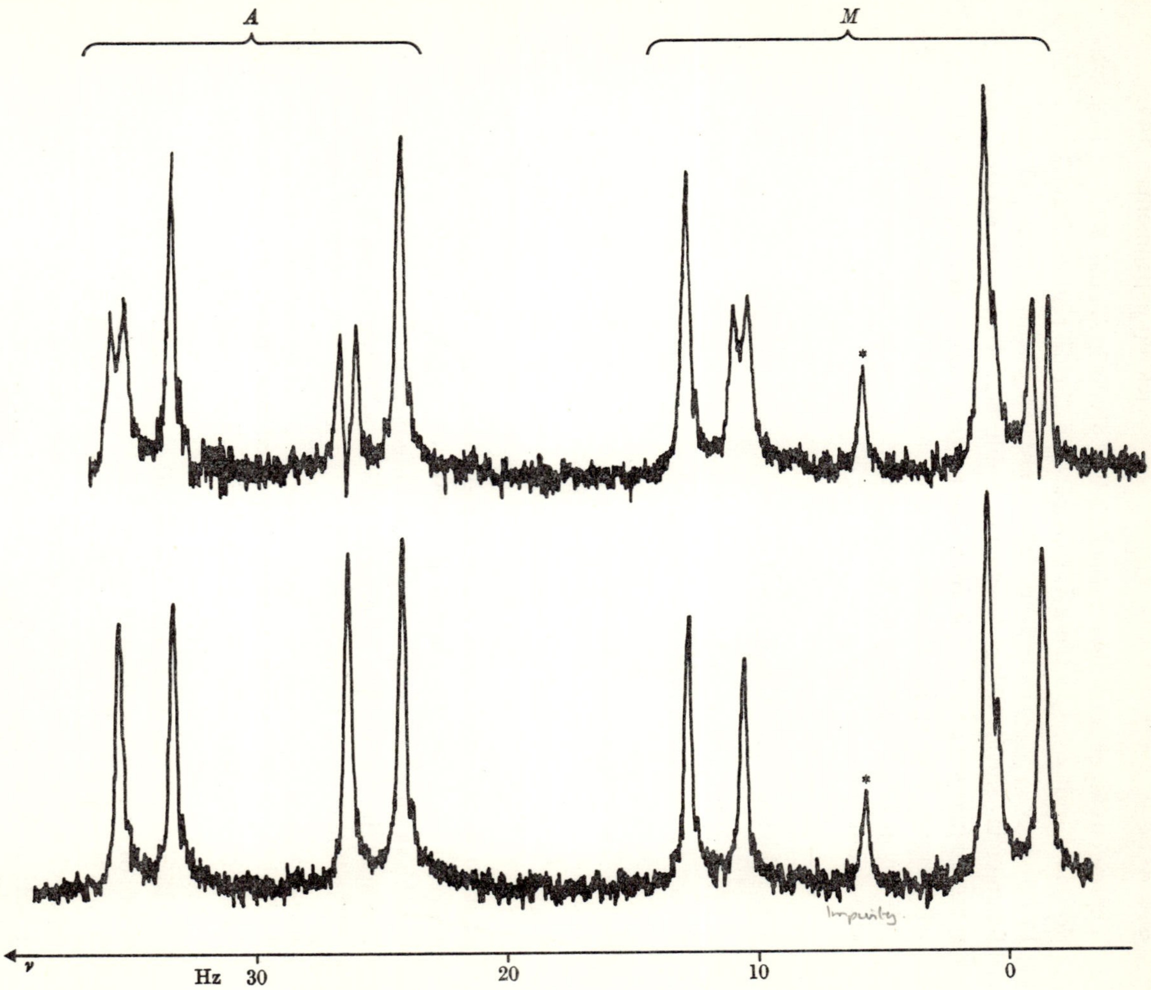

Fig. 4-15 Tickling experiment on an AMX system, $CCl_3CH_AClCH_MH_XCl$. The single resonance spectrum is shown below a double resonance spectrum in which the lowest frequency X line is irradiated. The line marked with an asterisk is due to an impurity. The AX and MX couplings can be seen to have opposite signs. A full analysis (see Fig. 4-16) shows that ${}^3J_{AM} = +2{\cdot}15$ Hz, ${}^3J_{AX} = +9{\cdot}4$ Hz, and ${}^2J_{MX} = -12{\cdot}0$ Hz. Note the better resolution of regressive transitions (see text). (We thank A. B. Dempster, K. Price and N. Sheppard for this spectrum.)

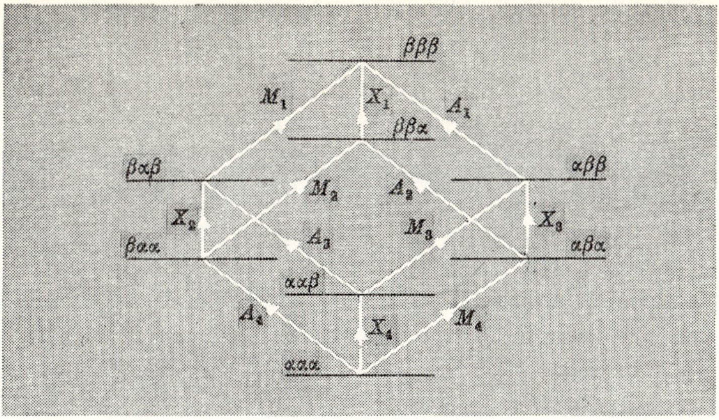

(a) A_1A_3	A_2A_4	M_1M_3	M_2M_4	X_1	X_3 X_2	X_4	all +
(b) A_3A_1	A_4A_2	M_3M_1	M_4M_2	X_1	X_3 X_2	X_4	J_{AM} −
(c) A_1A_3	A_2A_4	M_2M_4	M_1M_3	X_2	X_4 X_1	X_3	J_{MX} −
(d) A_2A_4	A_1A_3	M_1M_3	M_2M_4	X_3	X_1 X_4	X_2	J_{AX} −

Fig. 4-16 Energy level diagram and single resonance spectrum of an AMX system in which $|J_{MX}| > |J_{AX}| > |J_{AM}|$. The spectral lines are identified with transitions on the energy level diagram (a) if all couplings are positive (b) J_{AM} only is negative (c) J_{MX} only is negative (d) J_{AX} only is negative. In Fig. 4-15 the line marked with an arrow is irradiated and the lines marked with asterisks are split. Assignment (c) is consistent with the observations.

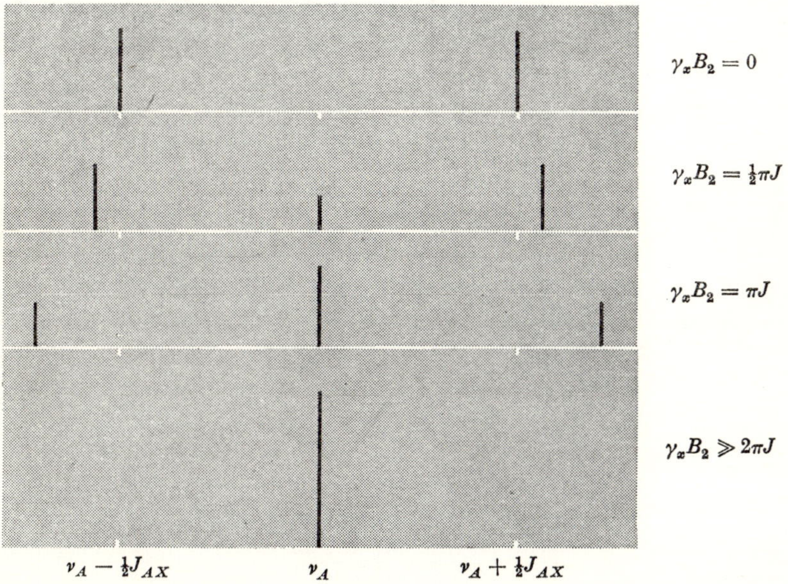

Fig. 4-17 Calculated double resonance spectra of the A part of an AX system when the perturbing field is applied at the centre of the X spectrum, showing the 'collapse' of the A doublet as B_2 is increased.

Obtained from the Hamiltonian sol^ns. Exact positions using calculated for various signs of coupling constant J_{ij}, giving the energy assignments below

The particular lines split may then be predicted from the energy diagram. only (c) corresponds with experiment.

that the two- and three-bond coupling constants in this molecule have opposite signs (Problem 4-6). It is not possible to determine absolute signs in a double resonance experiment.

In Fig. 4-15 some of the splittings are better resolved than others. In fact, both lines are split by an amount $\gamma_A B_2/2\pi$, but the effect of field inhomogeneity is that progressive pairs of transitions such as (M_2X_1), in which the three levels concerned have different values of the total z-spin quantum number m_T, are broader and therefore less well resolved than regressive transitions such as (M_2X_2), in which two of the three levels share an m_T value. This fact is often an aid to line assignment.

4-13 Line collapse and spin decoupling

Figure 4-17 shows the predicted double resonance spectra for the AX system as a perturbing field of increasing strength is applied at ν_X,

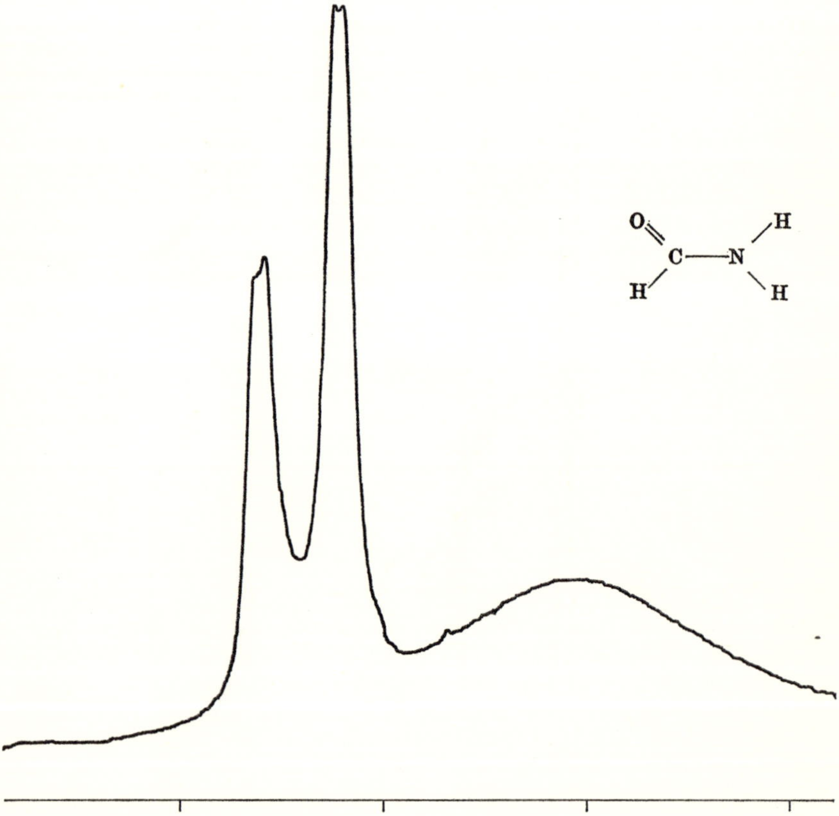

(a)

Fig. 4-18 ^{14}N Decoupling. (a) The normal 60 MHz ^{1}H spectrum of formamide showing considerable quadrupole broadening due to the ^{14}N nucleus.

the centre of the X doublet. A new line of increasing intensity appears at the centre of the doublet, while the doublet lines move apart and lose intensity. The final spectrum consists of a single line at ν_A. This phenomenon is known as line collapse, and needs perturbing field intensities of the same order as the original doublet splitting. This is considerably greater than for tickling experiments; also the point of irradiation does not coincide with a transition. The coupling no longer affects the spectrum, as the effective field for

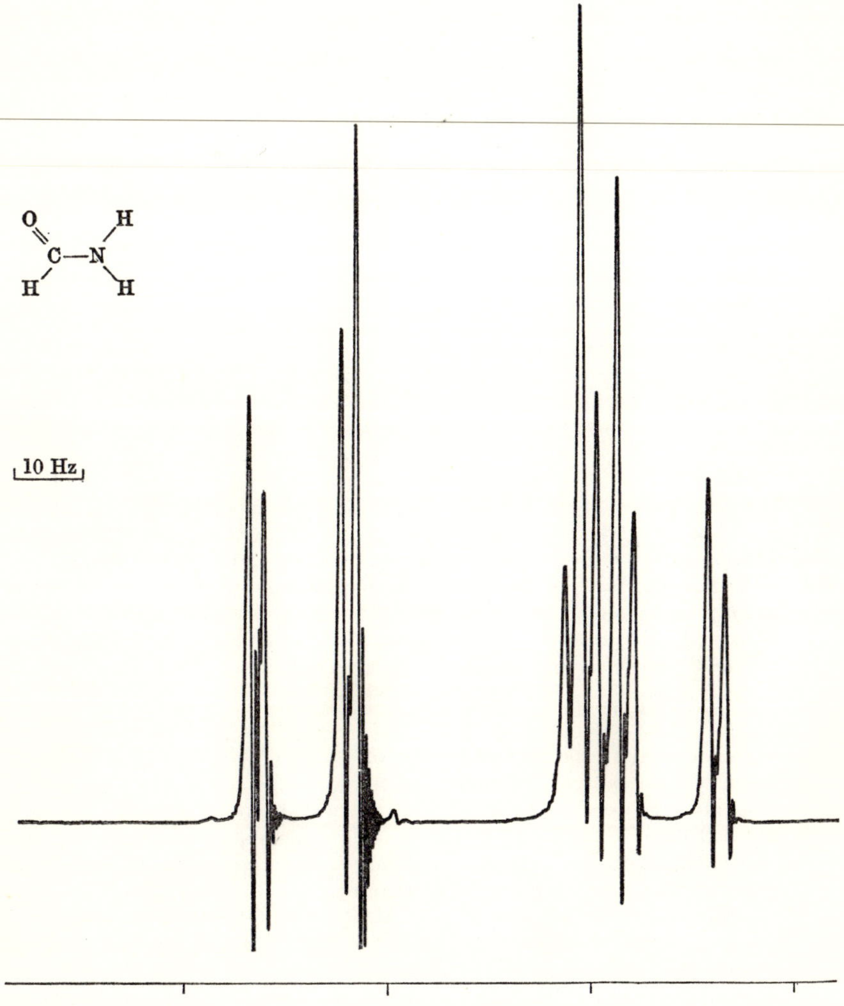

(b)

Fig. 4-18(b) The ^{1}H spectrum of formamide with ^{14}N decoupling at 4·33410 MHz. This is an ABC spectrum showing that rotation about the CN bond is slow at this temperature (20°C) so that protons 1 and 2 are non-equivalent. Full analysis of the spectrum yields $J_{12} = +2{\cdot}8$ Hz, $J_{13} = +18{\cdot}1$ Hz, and $J_{23} = +3{\cdot}3$ Hz. (By courtesy of JEOLCO).

the X nuclei is perpendicular to that for the A nuclei and $\langle \hat{\mathbf{I}}_A \cdot \hat{\mathbf{I}}_X \rangle$ is zero.

This technique can be extended to systems of more than two spins, and is useful for simplifying complex spectra by completely removing the effect of one nucleus. This is illustrated in Fig. 4-18.

Problems

4-1 Show that a nucleus of spin $\frac{1}{2}$ at thermal equilibrium at temperature T°K in a field B has a probability of

$$2 \exp - (\gamma\hbar B/2kT) [\cosh (\gamma\hbar B/2kT)]^{-1}$$

of being in the state α. Calculate the difference in the population of the two states of ^{13}C in a field of 3 T at 27°C.

4-2 Show that the equilibrium magnetization of a sample containing N nuclei of spin I can be written

$$M = \sum_{m=-I}^{I} n_m m\gamma\hbar$$

where n_m is the number of nuclei in the state $|m,I\rangle$. Show that at sufficiently high temperatures

$$\mathbf{M} = \tfrac{1}{3}\gamma^2\hbar^2 I(I + 1)\mathbf{B}/kT$$

4-3 Draw a diagram to illustrate approximately the variation in the direction and size of the magnetic moment in rotating axes under steady state conditions as the applied frequency is increased through resonance. Assume $T_1 = T_2$. From the variation in u and v in your diagram deduce the approximate forms of the absorption and dispersion spectra.

(Ans. For diagram, see the book by Pople, Schneider, and Bernstein, Fig. 3-4, page 38; note our definition of v has the opposite sign.)

4-4 Calculate the form of the quadrupole spectrum of the ^{35}Cl nucleus in an axially symmetric environment. Pure quadrupole transitions are observed at 25·8 MHz in K_2PtCl_6 and at 54·248 MHz in Cl_2. Calculate the field gradients and comment on the variation.

$$(Q = -0{\cdot}07894 \times 10^{-28}\,\text{m}^2)$$

4-5 Write down the equations of motion of v_A, u_A and v_B, u_B, the magnetizations of two samples of nuclei with resonance frequencies ν_A and ν_B respectively, referred to a set of axes rotating at frequency ν with the applied field. Suppose that a sample contains equal numbers of A and B nuclei, which can exchange at a rate k nuclei per second. Modify the equations to take exchange into account.

Assuming that $M_{Az} = M_{Bz} = M_0$, and that a steady state has been reached, solve the equations for $(v_A + v_B)$. This is the line shape of the system under conditions of exchange.

4-6 Assign the lines in Fig. 4-16 (an AMX spectrum) if (a) all couplings are negative and (b) J_{MX} only is positive, and find which assignment is consistent with the double resonance spectrum, Fig. 4-15. Verify that changing the signs of all the coupling constants does not alter the double resonance spectra. Deduce that irradiating an AMX spectrum in the M region and observing in the X region will only give the relative signs of J_{AM} and J_{AX}.

4-7 If $\psi(x, y, z)$ is a wave function referred to a static set of axes and $\psi'(x', y', z')$ is the same function referred to a set of rotating axes then

$$\psi' = \exp(i\omega \hat{I}_z t)\psi$$

when the axes are rotating with angular velocity ω about the z direction.

Show that the function $\hat{A}\psi$ becomes $\hat{A}'\psi'$ in the rotating axes where

$$\hat{A}' = \exp(i\omega \hat{I}_z t)\hat{A} \exp(-i\omega \hat{I}_z t)$$

show that

$$\langle m| \hat{A} |n\rangle = \langle m'| \hat{A}' |n'\rangle$$

and that the time-dependent Schrödinger equation becomes

$$i\hbar \frac{d\psi'}{dt} = (-\omega\hbar \hat{I}_z + \hat{\mathscr{H}}')\psi'$$

Hence deduce Eqn. (4-23) for the effective double resonance Hamiltonian.

Suggestions for further reading

Nuclear resonance

Comprehensive text books on this subject are Pople, Schneider, and Bernstein and the more recent Emsley, Feeney, and Sutcliffe. These deal with both principles and applications in more detail than this book. Carrington and McLachlan cover theoretical aspects of both electron and nuclear resonance and provide a useful link between these subjects. Hecht is another useful book on nuclear resonance.

There are three review series devoted to magnetic resonance which provide articles of varying sophistication and specialization. Other review series occasionally contain useful articles on the applications of nuclear resonance to various branches of chemistry. We list a few selected reviews.

Books

POPLE, J. A., W. G. SCHNEIDER, and H. J. BERNSTEIN. *High-resolution nuclear magnetic resonance.* McGraw-Hill, New York, 1959.

EMSLEY, J. W., J. FEENEY, and L. H. SUTCLIFFE. *High-resolution nuclear magnetic resonance spectroscopy* (2 volumes). Pergamon Press, Oxford, 1965.

CARRINGTON, A. and A. D. MCLACHLAN. *Introduction to magnetic resonance with applications to chemistry and chemical physics.* Harper and Row, New York, 1967.

SLICHTER, C. P. *Principles of magnetic resonance.* Harper and Row, New York, 1963.

JACKMAN, L. M. *Applications of nuclear resonance spectroscopy in Organic Chemistry.* Pergamon Press, Oxford, 1959.

HECHT, H. G. *Magnetic resonance spectroscopy,* John Wiley, New York, 1967.

Review series

Progress in nuclear magnetic resonance spectroscopy (Eds. J. W. EMSLEY, J. FEENEY and L. H. SUTCLIFFE). Pergamon Press, Oxford.

Advances in magnetic resonance (Ed. J. S. WAUGH). Academic Press, New York.

Annual review of NMR spectroscopy (Ed. E. F. MOONEY). Academic Press, New York.

Individual reviews

BARFIELD, M. and D. GRANT. Theory of coupling constants. *Adv. Mag. Res.* **1**., 149, 1965.

HOFFMAN, R. A. and S. FORRSEN. *Prog. Nucl. Mag. Res. Spect.* **1**., 15, 1966.

KUBO, M. and D. NAKAMURA. Chemical applications of quadrupole resonance. *Advan. Inorg. Radiochem.*, **8**, 257, 1966.

BOVEY, F. A. Application of NMR to polymer chemistry, in *Encyclopaedia of Polymer Science*, Vol. 9, pp. 356–396. John Wiley, New York, 1968.

Quantum mechanics

Matthews provides a good introduction to this subject using operators, which is particularly relevant as background reading for Chapter 2. There are many other text books such as Schiff which cover both normal and time-dependent perturbation theory. The calculation of molecular wave functions is treated by Murrell, Kettle and Tedder. Blin-Stoyle discusses the spins and magnetic moments of nuclei in terms of nuclear structure.

MATTHEWS, P. T. *Introduction to quantum mechanics.* 2nd edn. McGraw-Hill, New York, 1968.

SCHIFF, L. I. *Quantum mechanics*, 2nd edn. McGraw-Hill, New York, 1955.

BLIN-STOYLE, R. J. *Theory of nuclear moments.* Oxford University Press, Oxford, 1957.

MURRELL, J. N., S. F. A. KETTLE, and J. M. TEDDER. *Valence theory.* John Wiley, London, 1965.

Other topics

Units and dimensions are discussed with clarity in the booklet by McGlashan. There are many textbooks of electricity and magnetism which provide background for Chapter 3. Kip uses rationalized MKSA units as we do. The applications of magnetic resonance to molecular beams is dealt with by Ramsey, who also discusses the form of the magnetic interactions in the molecular Hamiltonian in some detail.

MCGLASHAN, M. L. *Physico-chemical quantities and units.* Royal Institute of Chemistry, 30 Russell Square, London, W.C.1, 1968.

KIP, A. F. *Fundamentals of electricity and magnetism*, McGraw-Hill, New York, 1962.

RAMSAY, N. F. *Molecular beams*, Oxford University Press, London, 1956.

Answers to problems

Chapter 1

1-1 $4{\cdot}581 \times 10^7$ rad T^{-1} s^{-1}

1-2 (a) 12·69 MHz (b) 11·42 MHz (c) 6·26 MHz

1-3 1·0 T, 2·47 T

1-6 0·796 and 8·856; 75·5, 80·0, 512·9, 517·4 Hz

1-7 520 Hz; —5·20 p.p.m.; 12·22 μT

Chapter 2

2-9 (a) 0 (b) $i/2[I(I+1) - m(m-1)]^{1/2}$ (c) 2
(d) $\frac{1}{2}[I(I+1) - m(m+1)]^{1/2}$ (e) $\sqrt{3}$ (f) 2

2-10 0; 0; $-\frac{1}{2}$; 0; $(c_1 + c_2 + c_3)$

Chapter 3

3-2 61° 31′

3-5 209 Hz

Chapter 4

4-4 One line with $\nu = e^2qQ/8\pi\varepsilon_0 h$

4-5
$$\frac{2\gamma B_1\, kM_0\,[(\bar{\nu} - \nu)^2 - (\nu_A - \nu)(\nu_B - \nu)]}{(2\pi)^2[(\nu_A - \nu)^2(\nu_B - \nu)^2 + (\bar{\nu} - \nu)^2(k/\pi)^2]}$$

Index

WADHAM COLLEGE
LIBRARY.